THE OFFICIAL POCKET EDIBLE PLANT SURVIVAL MANUAL

Robert W. Pelton

"The Official Pocket Edible Plant Survival Manual"
Robert W. Pelton © 1998

First Printing 1998

ISBN 0-9658219-3-5

Library of Congress Catalog Card Number
97-094548

Published By:

Freedom & Liberty Foundation

CONTENTS

PART I

INTRODUCTION TO EDIBLE PLANTS AND SURVIVAL

1

THERE'S NO NEED TO STARVE!

It doesn't take a genius to understand the one most important aspect of surviving. It's having access to drinkable water! Without water a person can't live more than three days.

The second most important thing is food! Men have been known to live more than a month without food. But there's absolutely no need for any person to be deprived of something to eat. Nature is and always has been a good and reliable provider. Everyone should know how to properly use her. Learn to live off the land. It really isn't that difficult.

STEEP: Put leaves or other plant parts in pot of boiling water. Take off stove and cover. Leave to steep for 10 to 20 minutes. When brewing tea, pour boiling water over crushed leaves in pot, cover, and let steep 5 to 15 minutes.

MAKING CANDY, FRITTERS, PIES AND MORE

CANDY MAKING:

1. Flowers and root pieces of various plants can be "candied" and eaten as a delicious sweet treat. The process is simple:
   a. Whip powdered sugar in egg whites.
   b. Dip flowers or root pieces into this meringue-like blend.

or

   a. Stir flowers in egg whites and then roll in sugar.
   b. Deep fry in pot of hot cooking oil until slightly browned.
   c. Take out and set on paper to drain and cool.
   d. Sprinkle with sugar-spice mixture.

or

   a. Take nut meats, roots, etc., and drop into pan of sugar syrup.*
   b. Simmer until plant parts or nut meats are nicely saturated.
   c. Take out of sugar syrup set aside to partially dry.

d. Roll pieces around in granulated sugar until covered.

* NOTE: Make sugar syrup by blending 2 cups sugar with every cup of water in pot.

COFFEE SUBSTITUTE MAKING:

a. Wash roots or seeds thoroughly or remove nuts from shells.
b. Roast slowly until brittle and dark brown.
c. Pulverize or grind roasted root, seeds or nuts.
d. Put 1 to 2 teaspoons grounds in pot boiling water.
e. Let simmer a short time.
f. Strain before serving.

NOTE: With chicory, for example, use 1-1/2 teaspoon grounds with each cup water.

FLOUR MAKING:

1. Flour can be made from variety of plant parts. It's obtainable from any edible plant by doing the following:
   a. Dry plant thoroughly over embers or in sun.
   b. Pulverize or grind plant to powder or meal.
   c. Sift out fiber.
   d. Store flour in air tight container.

2. Flour made from richly flavored cattail pollen is one of best known. Here's how it's done:
   a. Gather starch-filled rootstocks (late fall -- early spring).
   b. Wash thoroughly and peel away outer covering.
   c. Crush starchy core in bucket of cold water.
   d. Strain out fibrous material.
   e. Let starch settle to bottom.
   f. Carefully pour off water.
   g. Refill bucket with cold water.
   h. Let starch settle.
   i. Pour off water.
   j. Starch can be used as flour at this point.
3. Flour made with acorns from other than white oak trees:
   a. Boil acorns 2 hours and pour off water.
   b. Soak acorns 12 hours in cold water.
   c. Change soak water twice daily for 4 days.
   d. Now pound acorns to paste.
   e. Spread paste thinly and allow to dry.
   f. Pound into flour.

FRITTER MAKING:

1. Some flowers are suitable for making tasty fritters. Included are dandelions, day lilies, elderberries and wisteria. Here's how:
   a. Make a mildly thick batter by blending flour with water.
   b. Dip fresh flowers in this batter.
   or
   a. Dip fresh flowers in egg whites.
   b. Roll dampened flower in flour or fine meal.
   then:
   c. Drop individual flowers in pot of hot cooking oil.
   d. Deep fry until lightly browned and crisp.
   e. Eat while hot or at least while still warm.

JELLY MAKING

1. Flowers contain no pectin, nor do some fruits, yet they can still be used to make jelly. Making jelly isn't difficult, but it is time consuming. Here's how it's done:
   a. Add quart water for every quart berries in large pot.
   b. Bring to boil and let simmer 15 minutes.
   c. Mash berries and let simmer another 10 minutes.

d. Strain through cheesecloth several times.

2. Now prepare an equal amount of a high pectin fruit (crabapples, for example) in identical manner.
   a. When finished, combine the two strained juices.
   b. Sweeten to taste.
   c. Heat again until liquid begins to jell.
   d. Pour into sterilized jars and seal.
3. Tasty jam is made same way jelly is made except fruit isn't strained out as with jelly.

PANCAKE MAKING:

1. This recipe uses flour made from ground or pulverized plantain seeds. Other flour made from variety of plant parts can be substituted:

Plantain flour	2 cups
Sugar	3 tablespoons
Baking powder	3 teaspoons
Salt	1/2 teaspoon
Milk	1 cup
Cooking oil	3 tablespoons
Eggs	2

2. Blend above ingredients. Add more flour or milk as needed to get proper consistency. Fruit, if available, can be stirred into batter. Cook on hot griddle or in lightly greased frying pan.

PICKLING:

1. To make liquid suitable for pickling, combine and bring to boil:

Water	1 cup
Vinegar	2 cups
Then stir in:	
Sugar	1/2 cup
Salt	1/4 cup
Ground mustard	1 tablespoon
Celery seed	1 tablespoon

Boil 10 to 15 minutes while constantly stirring.

PIE MAKING:

Blend following ingredients:

Berries, grapes, etc.	4 cups
Apples	2 cups

Then stir in:

Maple syrup or brown sugar	4 tablespoons

Cinnamon	1 tablespoon
Raw eggs	2
Lemon juice	1 tablespoon

Pour mixture in pie shell. Cover with top crust. If no shell or crust available, just pour mixture in pie pan. Bake at 325 for 20 to 30 minutes.

PUREE MAKING (thick soup):

a. Put 2 cups leaves (or other plant parts) in pot of water.
b. Bring to boil and cook to pulp.
c. Run through sieve.
d. Stir in 2 tablespoons butter or margarine.
e. Simmer long enough to melt butter.
f. Stir in 1/4 cup milk.
g. Season to taste with salt and pepper.

SYRUP MAKING WITH FLOWERS:

a. Put 6 cups flowers in bowl.
b. Cover flowers with boiling water.
c. Let seep overnight.
d. Strain off flowers and throw away.

e. Stir in 2 cups sugar and bring mixture to boil.
f. Continue boiling until liquid starts thickening.
g. Use syrup for pancakes or whatever. Pour what is left in covered jars and store in cool place.

Or make syrup with rose hips:
a. Put quantity of rose hips in pot.
b. Add boiling water to cover.
c. Boil until hips become soft and liquid begins thickening.

SYRUP MAKING FROM TREE SAP:
a. Put quantity of sap into large cooking pot.
b. Set over fire and continuously stir.
c. Spoon off scum as it rises to top.
d. Don't allow sap to boil over or burn bottom of pot.
e. Bring to boil.
f. Let simmer until sap thickens and turns to clear amber syrup.
g. Continue simmering until teaspoon of syrup forms soft ball when held under cold water.
h. Remove from fire at this point.

SUGAR MAKING FROM TREE SAP:

a. Continue to simmer when sap begins to thicken and tastes sweet.
b. Cook until syrup starts crystallizing.
c. Syrup will eventually turn to sugar.

PART II

EDIBLE PLANTS -- WHAT PARTS TO USE AND HOW

3

FRUITS AND BERRIES FOUND IN THE WILDS

Edible fruit is plentiful in nature and it supplies great food in a survival situation. You're no doubt already aware of many of the wild fruits and berries in the United States. However, to refresh your memory, all the following are readily available, easy to find and are meticulously covered in this chapter.

Blackberry	Grapes	Blueberry	Mulberry
Crabapple	Persimmon	Elderberry	Rhubarb
Serviceberry	Strawberry		

BLACKBERRY (BRAMBLE)

BERRIES:

a. Eat fresh berries alone or in milk for nourishing quick snack.
b. Stir fresh berries into pancake or muffin batter.
c. Cook berries to make jelly or jam (Chapter 2).
d. Set aside boiled berry juice to ferment for making vinegar.

e. Crush berries and cook until juice thickens. Use for pancake syrup or fruit sauce.
f. Stew berries to make dessert or pie (Chapter 2).

LEAVES:

a. Use uncooked leaves as a salad green.
b. Cook (boil or steam) as vegetable green.
c. Add liberally to soups and stews.

YOUNG SHOOTS:

a. Eat shoots raw as snack or add to green salad.
b. Peel shoots, boil 15 minutes and eat like asparagus.

FLOWERS:

a. Eat raw as a snack or add to green salad.
b. Use for making fritters (Chapter 2).

NOTE:

1. Berries dried thoroughly can be stored indefinitely.
2. Blackberries are extremely high in Vitamin C.
3. Huckleberries and red raspberries are prepared in the same manner for eating as blackberries.

BLUEBERRY

BERRIES

a. Eat fresh berries alone or in milk for quick snack.
b. Stir fresh berries into pancake or muffin batter.
c. Boil berries to make jelly or jam (Chapter 2).
d. Set aside boiled berry juice to ferment for making vinegar.
e. Crush berries and cook until juice thickens. Use on pancakes and icecream.
f. Stew berries to make wholesome dessert.

LEAVES:

a. Use raw leaves as a salad green.
b. Cook (boil or steam) as a vegetable green.
c. Add liberally to soups and stews.

SHOOTS:

a. Eat young shoots raw as snack or add to green salad.
b. Peel and boil young shoots 15 minutes and eat like asparagus.

FLOWERS:

a. Eat raw as a snack or add to salads.
b. Use to make fritters (Chapter 2).

NOTE:

1. Berries dried in sun or over embers can be stored indefinitely.
2. Blueberries are extremely high in vitamin C.
3. Best time to find blueberries is early summer through early autumn.

CRABAPPLES, WILD

a. Eat entire crabapple raw.
b. Boil until soft, sweeten to taste and eat as dessert.
c. Steam crabapple and bake in pies (Chapter 2).
d. Crush cooked or uncooked crabapple and eat as applesauce.
e. Make into crabapple jelly (Chapter 2).
f. Fry slices of crabapple (Chapter 2).
g. Saute crabapple slices (Chapter 2).
h. Use pieces of crabapple to make fritters (Chapter 2).

LEAVES:

a. Boil until tender and eat as vegetable green.
b. Steam until tender and eat as vegetable green.
c. Saute in lightly greased frying pan (Chapter 2).

NOTE:

1. Crabapples sliced and dried in sun or over embers can be stored indefinitely.
2. Crabapples are sour and hard unless tree ripened.
3. Save liquid used for cooking plant parts and drink or use for soups and stews.

COOKING POINTERS:

1. Crabapples are almost always tart. Stew with honey, sugar or maple syrup for best results.
2. Crabapples contain abundant pectin for making jelly and jam. Use with other fruits containing no pectin.

ELDERBERRY

a. Eat fresh ripe berries for quick snack. They're edible but smell bad and don't taste very appetizing.
b. Stir fresh ripe berries into pancake or muffin batter.
c. Cook berries to make jelly and jam (Chapter 2).
d. Set aside boiled berry juice to ferment for making vinegar.
e. Crush berries and cook until thickened. Use for pancake syrup or fruit sauce.

f. Mash fresh or cooked ripe berries for a nice drink.
g. Stew berries to make dessert or pie filling (Chapter 2).

FLOWERS:

a. Make fritters with elderberry flowers (Chapter 2).
b. Soak flower clusters in water with lemon slices to make tasty elderberry lemonade rich in vitamin C.

NOTE:

1. Berries dried in sun or over embers can be stored indefinitely. 2. Drying uncooked berries takes away bad odor and taste.
2. Gather flower clusters in late spring for best fritters.
3. Elderberries have no pectin for making jelly and jam. Some other fruit (apples, etc.) must be used with them.
4. Berries high in vitamin C, A, potassium and calcium.
5. Vomiting, diarrhea and nausea is the consequence of eating unripe berries, leaves, roots and stem of this plant.
6. Save cooking liquid and drink or use for soups and stews.

GRAPES, WILD

GRAPES:

a. Eat ripe grapes uncooked as a snack.
b. Add ripe grapes to a salad along with the leaves.
c. Boil down to make jelly and jam (Chapter 2).
d. Use to make pie (Chapter 2).
e. Make raisins by drying grapes in sun for 3 days or more.

LEAVES:

f. Steam or boil 10 to 15 minutes and eat as vegetable.
g. Saute shredded leaves until lightly browned.
h. Stuff steamed larger leaves with grapes and other fruit.

TENDRILS:

a. Eat raw tendrils for quick snack or use in green salad.

NOTE:

1. Best to get leaves when tenderest during early summer.
2. Grapes are loaded with pectin and energy-giving sugar.
3. Drinkable water can be extracted from a grapevine.
4. Save cooking liquid and drink or use for soup or stew.

COOKING POINTERS:

a. Take large leaves and lightly boil or steam to soften.
b. Stuff with meat, vegetables, rice, etc.
c. Roll tightly up and put in colander.
d. Put colander in pot and let sit just above water.
e. Let steam until leaves are limp eat while hot.

MULBERRY, RED

BERRIES:

a. Eat ripe berries alone or in milk for nourishing quick snack.
b. Stir ripe berries into pancake or muffin batter.
c. Stew berries to make jelly or jam (Chapter 2).
d. Set aside boiled juice to ferment for making vinegar.
e. Crush berries and cook until juice thickens. Use for pancake syrup or fruit sauce.
f. Mash cooked berries thoroughly and use as fruit drink.
g. Stew berries to make dessert or pie (Chapter 2).

LEAVES:

a. Use uncooked leaves as a salad green.
b. Cook (boil or steam) and use as vegetable green.
c. Add leaves liberally to soups and stems.

SHOOTS:

a. Steam young shoots and eat as snack.
b. Peel young shoots, boil, and eat like asparagus.

NOTE:

1. Berries dried in sun or over embers can be stored indefinitely.
2. Don't eat raw shoots or unripe berries as they are mildly hallucinogenic.
3. Pectin is needed to make jelly and jam.
4. Pick shoots in the spring when they are tenderest.
5. Save cooking liquid and drink or use for soups and stews.

PERSIMMON

FRUIT:

a. Eat the sweet, plum-like fresh fruit raw for snack.
b. Strain seeds from pulp to make jelly and jam (Chapter 2).

c. Bake strained fruit pulp to make tasty pudding.
d. Make pie with strained pulp (Chapter 2).
e. Add fruit pulp to nut bread and muffins, etc.
f. Mash fruit and use as nourishing drink.

NOTE:

1. Fruit ripens in early fall.
2. Fully ripened fruit tastes not unlike a date.
3. Fruit is extremely sour until after first frost.
4. Fruit richly sweet when soft with wrinkled skin.
5. Save cooking liquid and drink or use for soup or stew.

RHUBARB, WILD

STALKS:

a. Eat stalks pieces raw as a rather sour snack.
b. Cut stalks into tiny pieces and add to green salad.
c. Stew or steam chunks until tender, add sugar to suit and eat as delightful fruit dessert.
d. Boil stalks until soft and mushy, add sugar to suit and use for pie filling.

e. Boil down until liquid in pot thickens and is ready for making jelly and jam (Chapter 2).
f. Steam stalks until tender, add sugar to suit and eat as dessert or use for pie filling.

LEAVES:

a. Boil leaves 20 to 25 minutes and eat as you would spinach.
b. Steam leaves until tender and eat as spinach.
c. Add liberally to soups and stews.

NOTE:

1. BEWARE: Rhubarb leaves are toxic if eaten raw.
2. Save cooking liquid to drink or use for soup and stew.

COOKING POINTERS:

1. Change water 2 or 3 times when cooking to get rid of some of sour taste.
2. When short of rhubarb, curly dock leaf stems are excellent substitution for pies, puddings and sauces.

SERVICEBERRY (JUNEBERRY)

BERRIES:

a. Eat fresh berries alone or with milk for nourishing snack.
b. Stir fresh berries into pancake or muffin batter.
c. Cook berries to make jelly or jam (Chapter 2).
d. Set aside boiled juice to ferment for making vinegar.
e. Crush, cook until juice thickens and use as pancake syrup.
f. Mash fresh berries thoroughly and use as fruit drink.
g. Stew berries to make dessert or pie (Chapter 2).

LEAVES:

a. Use uncooked leaves as salad green.
b. Cook (boil or steam) and eat as a green vegetable.
c. Add liberally to soups and stews.

SHOOTS:

a. Eat young shoots raw as snack or add to green salad.
b. Peel young shoots, boil 15 minutes, and eat like asparagus.

FLOWERS:

a. Eat raw as tasty snack or add to green salads.
b. Use for making fritters (Chapter 2).

NOTE:

1. Berries dried in sun or over embers can be stored indefinitely.
2. Serviceberries are small dark blue, apple-like fruit with an almond-like flavor.
3. Save cooking liquid and drink or use for soup and stew.

STRAWBERRY, WILD

PLANT:

a. Steam or boil entire plant and eat.

BERRIES:

a. Eat fresh berries alone or with milk for nourishing snack.
b. Stir fresh berries into pancake or muffin batter.
c. Cook berries to make jelly or jam (Chapter 2).
d. Set aside boiled juice to ferment for making vinegar.
e. Crush, cook until juice thickens and use for pancake syrup.
f. Stew berries to make dessert or pie filling (Chapter 2).

LEAVES:

a. Use uncooked leaves in green salads.
b. Cook (boil or steam) and eat as green vegetable.
c. Add liberally to soups and stews.

SHOOTS:

a. Eat young shoots raw as snack or in green salad.
b. Peel young shoots, boil 15 minutes and eat like asparagus.

FLOWERS:

a. Eat raw as tasty snack or add to green salads.
b. Use for making fritters (Chapter 2).

NOTE:

1. Berries dried in sun or over embers can be stored indefinitely.
2. Berries and leaves are extremely high in Vitamin C.
3. Save cooking liquid and drink or use for soup and stew.

4

WILD PLANTS USED AS VEGETABLES

Various authorities estimate there are approximately 300,000 plants (those that have been classified) on the surface of the earth. Of these, 120,000 varieties have been determined to be edible.

A person should know what edible plants to look for when in the wilderness. He or she should also be able to properly identify these plants and to properly prepare them for eating.

Such an individual will undoubtedly find enough plant food out there to keep alive over an extended period of time. And he or she may even surprise themselves with a delicious meal.

ALFALFA

LEAVES:

a. Use uncooked leaves as salad greens.
b. Boil until tender and serve as vegetable green.
c. Steam 15 minutes and serve as vegetable green.

d. Add liberally to soups and stews.
e. Pulverize dried leaves and add to breakfast cereal.
f. Crush dried leaves into powder and steep in hot water for nutritious drink.

SPROUTED SEEDS:

a. Eat alone as tasty snack.
b. Use as nourishing addition to green salads.
c. Excellent on sandwiches with meat, cheese, etc.
d. Stir into soups and stews as nutritional supplement.

FLOWERS:

a. Eat raw as snack or add to green salads.
b. Use for making delicious fritters (Chapter 2).
c. Saute flowers until lightly browned (Chapter 2).
d. Crush dried flowers and add to breakfast cereals.
e. Crush dried flowers and add to soups and stews.

NOTE:

1. Alfalfa seeds are highly nutritious. They contain chlorophyll, amino acids, vitamins A, C, D, K, etc.
2. Drink nourishing liquid used to cook plant parts.

AMARANTH. GREEN

ENTIRE PLANT IS EDIBLE:

a. Cut up and use raw for salad.
b. Boil or steam until tender and eat.
c. Put in soups and stews.

LEAVES:

a. Use uncooked leaves as salad greens.
b. Boil or steam until tender and eat.
c. Add liberally to soups and stews.
d. Grind or pulverize dried leaves to make flour for baking.

SEEDS:

a. Seeds can be eaten raw or dried as cereal.
b. Seeds can be used raw or dried for baking.
c. Seeds can be crushed or ground and eaten as mush for breakfast, lunch or supper.
d. Shiny black seeds, harvested in the fall, can be parched (scorched or burned slightly) and then ground to make flour.

NOTE:

1. This dark flour is good for making waffles, pancakes and for baking homemade bread.
2. This plant has a delicate flavor unlike any other green.
3. When greens taste dry and coarse, add meat drippings.
4. This plant is high in vegetable protein.
5. Drink nourishing liquid used to cook plant parts.

ASPARAGUS, WILD

YOUNG STALKS:

a. Cut into small pieces and eat raw as snack.
b. Cut up raw stems and add to green salads.
c. Boil until tender and eat as vegetable green.
d. Steam about 10 minutes or until tender and eat.
e. Cut up and add liberally to soups and stews.
f. Saute cut up pieces (Chapter 2).
g. Deep fry cut up pieces as you would French fries.
h. Use cut pieces to make fritters (Chapter 2).
i. Use to make a delicious cream soup.
j. Cut into pieces, stir in eggs and fry as an omelet.

NOTE:

1. Eat only young shoots or stalks as you would cultivated asparagus. Older stalks are toxic.
2. This is one of the finest early spring vegetables.
3. Wild asparagus is best when eaten hot with butter.
4. Wild asparagus is similar to that purchased in stores.
5. Drink nourishing liquid used to cook this plant.
6. Young wild asparagus shoots are difficult to find.

COOKING POINTERS:

1. Use wild asparagus in any recipe calling for asparagus.
2. Wash and peel rough covering near base of stalk before cooking.
3. Tie stalks in small bundles and put into pot of boiling water.

BULRUSH

ROOTS:

a. Boil or steam until tender and eat as potatoes.
b. Boil into a gruel, let dry and use for flour.
c. Bake until tender and eat as baked potato.
d. Roast until done and eat as potato.

e. Roast several hours until dry and pound into flour.
f. Slice up and fry as you would potatoes.
g. Cut into pieces and eat raw in salads with other plant parts.

LEAVES:

a. Tear up raw young leaves and use in green salads.
b. Cook (boil or steam) young leaves and eat as vegetable.
c. Add liberally to soups and stews.

FLOWERS:

a. Dry and grind to make flour.
b. Use for making delicious fritters (Chapter 2).
c. Add liberally to soups and stews.
d. Used raw in salads.

BUDS:

a. Add uncooked spring buds to salads.
b. Boil or stew spring buds and eat as vegetable.
c. Add liberally to soups and stews.

POLLEN:

a. Pulverize (pound) to make flour suitable for baking.

WHITE STEM BASE:

a. Cut up and eat raw in salads with other plant parts.
b. Cook (boil or steam) and eat as vegetable.

UNDERGROUND STEMS:

a. Roast and eat like potatoes.

SEEDS:

a. Gather in the fall and grind into meal or flour.
b. Roast seeds and eat as quick snack.
c. Roast seeds and add to salads.

EARLY OR YOUNG SHOOTS:

a. Eat raw by themselves or use in green salad.
b. Boil or steam and eat as a vegetable.
c. Add liberally to soups and stews.
d. Saute shoots until lightly browned (Chapter 2).
e. Slice and fry as you would potatoes.
f. Roast shoots and eat as vegetable.

NOTE:

1. Drink nourishing liquid used to cook plant parts.
2. Bulrush roots are loaded with starch and sugar.
3. Flour made from bulrush seeds tastes nutty and sweet.
4. Toast seeds slightly to get rid of insects.
5. Young firm roots are a good source of sugar.

COOKING POINTER:

1. Bruise some young root pieces and cover with boiling water. Boil until water is almost gone to make sweet syrup.

BURDOCK

ENTIRE PLANT IS EDIBLE:

a. Boil or steam until tender and eat as green vegetable.
b. Use in soups and stews.

LEAVES:

a. Eat raw as quick snack.
b. Use uncooked leaves as salad greens.
c. Add liberally to soups and stews.

d. Boil or steam in 2 or 3 changes of water until tender.
e. Saute leaves (Chapter 2).

LEAF STEMS:

a. Boil until tender and save broth to make nice tasting soup.
b. Boil 10 to 15 minutes and eat as asparagus.
c. Steam until tender and eat as asparagus.
d. Stir fry in lightly greased pan until lightly browned.

STEM PITH:

a. Eat raw and salted as a nice snack.
b. Add to green salads for taste treat.
c. Boil or steam until tender and eat as vegetable.
d. Bake until tender and eat as baked vegetable.

FLOWER STALKS:

a. Boil core, mash, make patties and fry as you would potatoes.
b. Peel bitter green skin from long plump flower stalk. Slice into pieces and fry, boil, steam, etc., as you would potatoes.
c. Peel rough stalk when flower heads start to form. Eat as you would celery, by itself or in green salads.

BUR:

a. Steam and eat whole bur if young and pliable.

ROOTS:

a. Boil root of young plant after peeling off tough outer skin. Slice into thin strips and serve with butter.
b. Boil or steam strips of roots as you would carrots.
c. Roast roots until nicely browned and serve with butter.
d. Bake root whole until done and eat with butter.

NOTE:

1. Drink nourishing liquid used to cook plant parts.
2. Burdock is high in vegetable protein.
3. Burdock root has a nice sweet taste.

COOKING POINTERS:

1. Boil roots in 2 or 3 changes of water to remove bitter taste.
2. Mash roots after boiling, make into patties and fry.
3. Dandelion, wild onions or leeks and wild carrots (Queen Anne's Lace) should be combined with burdock in beef or chicken broth and simmered until everything is tender.

4. Flower stalks can be made into candy by simmering them in sugar syrup or maple syrup.
5. Barely cover peeled and sliced root with water. Add 1/4 teaspoon baking soda. Simmer 20 minutes. Drain water and cover root with fresh. Simmer until tender for delicious vegetable dish.

CAMAS LILY (WILD HYACINTH)

BULBS:

a. Boil 20 to 30 minutes and eat like potatoes (Chapter 2).
b. Slice boiled bulbs, dry in sun and store for future use.
c. Steam until tender throughout and eat like potatoes.
d. Bake in foil at 325 degrees for 45 minutes and eat.
e. Bake wrapped bulbs in pit 2 to 3 days. When finished baking, bulbs will be dark and sugary.
f. Cut up bulbs and add to soups and stews.
g. Slice raw bulbs and fry as potatoes.
h. Deep fry cut up pieces like French fries.
i. Saute bulb slices until lightly browned (Chapter 2).
j. Use bulb slices to make fritters (Chapter 2).

NOTE:

1. Bulbs edible all year. They taste good and are nutritious.
2. Bulb may be confused with death camas if collected before or after they bloom. Only collect bulbs when plant is in bloom.
3. Bulbs pleasant tasting with gum-like texture.
4. Bulbs extremely high in sugar.

COOKING POINTERS:

a. Indians had special method for cooking bulbs:
b. They were baked in pits until done.
c. Bulbs were then set aside to dry in sun.
d. This method of cooking bulbs got rid of gummy problem.

CATTAIL

a. Boil or steam until tender and eat as vegetable.
b. Use in soups and stews.

ROOTS:

a. Eat delicious and nutritious starchy root raw for quick energy.
b. Dry roots and pound or grind into flour.

c. Peel rootstalk, grate inner portion, boil and eat.
d. Peel raw rootstalk and put tender core in salads.
e. Roast roots in embers of fire and eat (Chapter 2).
f. Bake roots until done and eat as baked potatoes.

LEAVES:

a. Boil until tender and eat like asparagus.
b. Steam for 15 minutes and eat as green vegetable.

EARLY SHOOTS:

a. Eat young, tender, cucumber-tasting shoots as raw snack.
b. Eat tender raw shoots in salads.
c. Boil young shoots and eat like asparagus.
d. Saute young shoots (Chapter 2).

POLLEN:

a. Mix yellow pollen with water, form loaf and steam as bread.
b. Blend protein rich pollen with regular flour and use for baking bread, biscuits etc.

FLOWER SPIKES OR HEADS:

a. Eat early green spikes raw as highly nourishing snack.

b. Put raw early green spikes in salads with greens.
c. Dry early spikes and grind or pound into flour.
d. Boil pollen spikes 10 minutes in lightly salted water and eat like corn on the cob.

STEM PITH

a. Peel stem, take out pith and eat it raw.
b. Boil or steam pith for 10 minutes and eat while hot.

NOTE:

1. Scrub root when pulled from water and peel before cooking.
2. Roots and lower stem portion of cattail are sweet tasting and loaded with carbohydrates.
3. Crush roots in cold water and remove starch. This starch can later be used to thicken soups, stews and gravies, etc.
4. Cattails are highly nutritious plants. The edible rootstalks contain about 46% starch and 11 % sugar.
5. When yellow pollen is ripe, bend stalk over and shake into bag.
6. Dry pollen over embers or in sun before storing for later use.
7. Cattails top the list of edible wild plants because of its versatility, widespread abundance and ease of availability.

COOKING POINTERS:

1. Add 1 cup honey to 5 cups pollen to make a high energy food.
2. Blend 1 cup pollen to 2 cups pancake mix.
3. Blend 1 cup pollen to 2 cups flour to make bread, etc.
4. If shoots are tough, simply drop into a pot of salted water and let simmer to restore their tenderness.

CHAMOMILE

ENTIRE PLANT IS EDIBLE:

a. Cut up and add to green salad.
b. Boil 10 to 12 minutes or until tender.
c. Steam until tender and eat as vegetable.
d. Put in soups and stews.
e. Saute in lightly greased frying pan (Chapter 2).

LEAVES:

a. Eat raw as crispy snack.
b. Use uncooked leaves as salad greens.
c. Boil until tender and eat as vegetable green.
d. Steam 15 minutes and eat as vegetable green.

e. Add liberally to soups and stews.
f. Saute in lightly greased frying pan (Chapter 2).

FLOWERS:

a. Eat raw by itself as tasty snack.
b. Add to green salads.
c. Saute flowers until lightly browned (Chapter 2).
d. Use for making delicious fritters (Chapter 2).

TWIGS:

a. Dry twigs out and use for tasty and crunchy treat.

NOTE:

1. Chamomile is very high in minerals.
2. Always drink leftover liquid when finished cooking.

CHICKWEED

ENTIRE PLANT IS EDIBLE:

a. Boil or steam 2 to 5 minutes and eat like spinach.
b. Put in soups and stews before stems get tough.
c. Young uncooked plants make outstanding salad greens.

d. Stew entire plant in pot with beef, chicken, or other meat.
e. Saute in lightly greased frying pan (Chapter 2).
f. Chop up plant, blend with eggs and make delicious omelet.

LEAVES:

a. Boil or steam 2 minutes or until tender and eat like spinach.
b. Use young uncooked leaves as salad greens.
c. Add liberally to soups and stews.
d. Stir into eggs and fry as omelet.

FLOWERS:

a. Put uncooked flowers in salads.
b. Add flowers to soups and stews.
c. Use for making delicious fritters (Chapter 2).
d. Stir flowers into eggs and make tasty omelet.

STEMS:

a. Add raw young stems to green salads.
b. Boil or steam tender stems 2 to 5 minutes and eat as vegetable.
c. Saute until slightly browned (Chapter 2).

d. Cut up stems and fry in butter.
e. Cut into small pieces, blend with eggs and make omelet.

NOTE:

1. Chickweed has bland taste. Try mixing with stronger-tasting greens such as mustard, dandelion, chicory or watercress.
2. Always save liquid from cooking and use for nutritious drink.
3. Young leaves and stems, when cooked, taste much like spinach.
4. Chickweed is highly nutritious and rich in Vitamin C.
5. Chickweed is tenderer than most other wild greens.

COOKING POINTERS:

1. Seeds obtained in the fall can be used in soups as a thickener.
2. To make unique pancakes:
   a. Drain and then blot 2 cups boiled or steamed leaves.
   b. Add enough leaves to pancake batter to get correct consistency. c. Fry and eat with butter and jam or jelly.

CHICORY, WILD

LEAVES:

a. Use tender, young, white parts of leaves for salads.
b. Boil or steam above ground parts of leaves until tender.
c. Saute leaves (Chapter 2).
d. Add liberally to soups and stews.

FLOWERS:

a. Eat raw by themselves as snack or add to salads.
b. Saute in lightly greased frying pan (Chapter 2).
c. Use to make delicious fritters (Chapter 2).

ROOTS:

a. Boil or steam carrot-like roots and eat when tender.
b. Grind or crush roasted roots to make rather bitter coffee substitute (Chap. 2).

NOTE:

1. Always drink nourishing liquid used for cooking.
2. When boiling leaves use just enough water to cover.
3. Chicory leaves, highly favorable, can be added to some of the more bland tasting greens such as chickweed and purslane.

4. Pick leaves early as they become bitter with age.
5. Leaves look like dandelion but are thicker and rougher.

COOKING POINTERS:

1. Changes of water one or more times while boiling might be required to get rid of objectionable bitterness of plants.

CLOVERS

ENTIRE PLANT IS EDIBLE:

a. Boil plant and eat as vegetable.
b. Steam plant 10 to 15 minutes and eat.
c. Put plant in soups and stews.

LEAVES:

a. Use uncooked leaves as salad greens.
b. Boil or steam 5 to 10 minutes and eat as green vegetable.
c. Add liberally to soups and stews.
d. Saute leaves (Chapter 2).
e. Dry leaves over embers or in sun and pulverize to make flour.
f. Crush dried leaves and add to breakfast cereal.

FLOWERS:

a. Eat blossoms or round flower heads raw as snack.
b. Put blossoms or raw flower heads in salads.
c. Boil blossoms until tender and eat as vegetable.
d. Steam blossoms 15 minutes and eat as vegetable.
e. Add liberally to soups and stews.
f. Saute blossoms (Chapter 2).
g. Use for making delicious fritters (Chapter 2).
h. Dry blossoms and grind or pound into flour.

STEMS:

a. Raw stems should be eaten raw as snack.
b. Raw stems should be cut up and added to green salads.
c. Cut into pieces, stir into eggs and make omelet.
d. Boil until tender and eat as vegetable.
e. Steam 15 minutes and eat as vegetable.

SEEDS:

a. Crush and boil to make mush.
b. Parch seeds for tasty snack.
c. Eat seeds as a cereal.

d. Dry seeds and grind or pound into nutritious flour.
e. Seeds should be added to soups and stews as flavoring.

NOTE:

1. Clover is very rich in protein.
2. Always drink nourishing liquid used for cooking.

COMFREY

ENTIRE PLANT IS EDIBLE:

a. Boil or steam until tender and eat as vegetable.
b. Put in soups and stews.

LEAVES:

a. Eat early leaves (March and April) uncooked in salads.
b. Boil or steam until tender and eat like spinach.
c. Add liberally to soups and stews.

STALKS:

a. Boil or steam until tender and eat as vegetable.
b. Cut up and add to soups and stews.

FLOWERS:

a. Eat raw as snack or add to green salads.
b. Boil or steam until done and eat as vegetable.
c. Add liberally to soups and stews.
d. Saute flowers until lightly browned (Chapter 2).

ROOTS:

a. Boil or steam until tender and eat with butter.
b. Bake or roast until done and ready to eat.
c. Add whole or cut up in soups and stews.

COOKING POINTERS:

1. Older leaves bitter but still edible. Boil in 3 water changes.
2. Hairiness eliminated somewhat when plant is cooked.
3. Drink nourishing liquid used for cooking plant parts.

DAISY, OX-EYE

LEAVES:

a. Chew raw leaves by themselves for tasty snack.
b. Put raw leaves in green salads.

c. Boil one minute and eat as green vegetable.
d. Steam lightly and eat as green vegetable.
e. Add liberally to soups and stews.

FLOWERS:

a. Eat raw by themselves for tasty snack.
b. Use as attractive yet nourishing addition to green salads.
c. Add liberally to soups and stews.
d. Blend flowers with batter for pancakes or muffins.
e. Use for making delicious fritters (Chapter 2).

STEMS:

a. Eat raw by themselves for quick snack.
b. Stir liberally in soups and stews.
c. Steam lightly and eat as vegetable.
d. Saute lightly (Chapter 2).
e. Cut into small pieces, stir into eggs and fry as an omelet.

NOTE:

1. Drink highly nutritious liquid left over from cooking.
2. Also save cooking liquid for making soups and stews.

DANDELION

ENTIRE PLANT IS EDIBLE:

a. Boil 15 to 20 minutes and eat like spinach.
b. Steam until tender and eat as vegetable green.
c. Use entire plant in soups and stews.

LEAVES:

a. Eat raw young leaves by themselves as nourishing snack.
b. Use uncooked young leaves with roots and flowers for salads.
c. Boil for 10 minutes and eat like spinach.
d. Steam until tender and eat like spinach.
e. Add liberally to soups and stews.
f. Add chopped leaves to eggs and fry for delicious omelet.

STALKS/STEMS:

a. Cut up stems and eat as quick snacks.
b. Add to green salads.
c. Boil 10 to 15 minutes and eat as vegetable.
d. Steam until tender and eat as a green vegetable.

e. Add liberally to soups and stews.
f. Cut into small pieces and saute (Chapter 2).

FLOWERS:

a. Eat by themselves after twisting green base from sweet petals.
b. Eat raw in salads with leaves and roots.
c. Boil buds for 10 minutes and eat with butter.
d. Saute in lightly greased frying pan (Chapter 2).
e. Make terrific fritters (Chapter 2).

ROOTS:

a. Eat raw, pleasant tasting roots by themselves as snack.
b. Cut up root and put in salads with leaves and flowers.
c. Slice and boil until tender as you would carrots.
d. Steam sliced root for 15 to 20 minutes or until tender.
e. Roast slowly until dark brown throughout. Pulverize or grind and brew exactly like coffee.
f. Saute slices of root (Chapter 2).
g. Fry sliced root as you would potatoes.
h. Thoroughly dry and pulverize root to make flour.

NOTE:

1. Dandelions are more nourishing than any plant found in gardens or the wilds.
2. Dandelions are richer in beta-carotene than carrots and are loaded with vitamin A, calcium and vegetable protein.
3. Dandelion leaves are sweeter in the fall after a few frosts.
4. Gather young leaves before buds and flowers appear. They are more tender and taste better. Leaves get bitter with age.

COOKING POINTERS:

1. Tough, bitter tasting older leaves can be boiled twice to get rid of their bitterness and then added to other cooked greens.
2. Add strong-flavored dandelion greens to bland tasting greens such as chickweed and purslane to make them more edible.
3. Save the liquid used for cooking and use as a soup base.

DAY LILY

UNDERGROUND TUBERS:

a. Cut up crisp raw tubers and eat as healthy snack.
b. Cut up crisp raw tubers and use in salads.
c. Boil 15 to 20 minutes as you would corn on the cob.

d. Steam until tender.
e. Slice and fry until browned as you would potatoes.
f. Cut into strips and deep fry like French fried potatoes.
g. Make tasty fritters (Chapter 2).

BUDS:

a. Boil unopened buds 5 minutes and eat like green beans.
b. Steam unopened buds until tender and eat like green beans.
c. Add fresh or dried buds liberally to soups and stews.
d. Deep fry fresh buds as you would French fried potatoes.
e. Use to make fritters (Chapter 2).

FLOWERS:

a. Add flowers to green salads.
b. Cook in stews to season.
c. Put in soups and simmer as a thickener.
d. Use fresh flowers to make delectable fritters (Chapter 2).

SHOOTS:

a. Boil young or early shoots and eat like asparagus (Chapter 2).
b. Steam young shoots until tender and eat like asparagus.

NOTE:

1. Buds and flowers can be dried in the sun and then set aside for later use.
2. Root tubers can be dug up all year round.
3. The day lily is a great source of food, but few people know about this easy-to-find plant.
4. For seasoning in soups and stews, fresh or dried flowers can be used as well as withered flowers still on the plant.
5. Uncooked flowers for use in salads have a strong taste.
6. Tubers make a remarkable tasting vegetable when boiled, steamed or fried.
7. Always save and drink nourishing liquid used for cooking.

COOKING POINTERS:

1. Melt cheese or butter over deep fried buds (Chapter 2).

EVENING PRIMROSE, COMMON

LEAVES:

a. Peel tender new leaves and use sparingly as flavorful addition to tossed salad.
b. Boil peeled young leaves 20 minutes or until tender.
c. Steam peeled young leaves until tender and eat as vegetable.

FLOWERS:

a. Put liberally into soups and stews.
b. Make delicious fritters (Chapter 2).
c. Make into a tasty candy (Chapter 2).

SEEDS:

a. Crush seeds, boil and skim oil from surface of water.

ROOTS:

a. Boil roots for at least 10 minutes and eat as vegetable.
b. Steam roots until tender and serve with butter.
c. Slice cooked roots and fry until browned (Chapter 2).
d. Cut boiled or steamed roots in strips and deep fry as you would French fries.
e. Slice roots and put liberally in soups and stews.
f. Boil roots until tender and then simmer in sugar syrup for 20 to 30 minutes or until candied.

NOTE:

1. Freshly picked new leaves taste slightly bitter after cooking but are palatable.

2. Fresh roots, usually quite peppery tasting, are milder if dug up in early spring or late fall.
3. Some evening primrose roots are very tasty -- somewhat nut-like, and extremely nutritious.

COOKING POINTERS:

1. Leaves should be boiled in 3 changes of water to get rid of some of the objectionable bitterness.
2. Always boil or steam root slices first before frying if the flavor is stronger than desired.
3. Taproots will sometimes have an objectionable strong peppery taste. Here's what to do about this problem:
   a. Peel taproots and boil 20 to 30 minutes in 3 water changes.
   b. Then let simmer 20 more minutes, serve with butter, vinegar and salt.

FIREWEED (PILEWORT)

YOUNG LEAVES:

a. Eat raw leaves as nutritious snack.
b. Use uncooked leaves for green salads.

c. Boil or steam until leaves can easily be pierced with fork.
d. Add to soups and stews as pot herb.

FLOWER BUDS:

a. Eat clusters of raw buds for a nourishing snack.
b. Add raw bud clusters to any tossed salad.
c. Generously add bud clusters to soups and stews.
d. Boil bud clusters 10 to 15 minutes and eat with butter.
e. Steam bud clusters until tender and serve as vegetable.

YOUNG SHOOTS

a. Eat raw young shoots as a tasty snack.
b. Use as nourishing addition to green salads.
c. Boil in salt water until tender and eat like asparagus.
d. Steam until tender and eat as vegetable.
e. Stir generously into soups and stews.

FLOWER STALKS:

a. Cut up and eat raw as snack.
b. Add to tossed or green salads.
c. Boil 19 to 15 minutes and eat as vegetable.

d. Steam until tender and eat as vegetable.
e. Add generously to soups and stews.

PITH FOUND IN STEMS WHEN PEELED:

a. Eat raw by itself as nourishing snack.
b. Add to tossed or green salads.
c. Put in boiling water until it becomes a thick soup.
d. Steam the pith until it becomes a thick soup.
e. Add as a thickener to soups and stews.

NOTE:

1. Fireweed is an extremely nutritious plant.
2. Far from the most tasty wild plant but it's still edible.
3. Has strong flavor. Must acquire a taste for this plant.
4. Young shoots should be collected in the spring.
5. Bud clusters should be picked before flowers have bloomed.
6. Young stems bear remarkable resemblance to asparagus.

GALINSOGA (QUICKWEED)

LEAVES:

a. Put uncooked leaves in salads.
b. Boil for 10 to 15 minutes and eat as vegetable.
c. Steam until tender and eat with butter, salt and pepper.
d. Add liberally to soups and stews.

FLOWERS:

a. Eat raw as a quick snack.
b. Use uncooked leaves in salads.
c. Saute flowers (Chapter 2).
d. Use to make delicious fritters (Chapter 2).

STEMS:

a. Boil for 15 minutes and eat as green vegetable.
b. Steam until tender and eat as green vegetable (Chapter 2).
c. Cut into pieces and add to soups and stews.
d. Deep fry cut up pieces like French fries.
e. Cut into small pieces, mix with eggs and make omelet.

NOTE:

1. Excellent dishes when served with butter or vinegar.
2. Serve with more flavorful greens to make up for bland flavor.
3. Save cooking liquid to drink or use for soups and stews.

GOOSEGRASS (CLEAVERS)

LEAVES:

a. Use Uncooked spring or early leaves in tossed salads.
b. Boil 10 to 15 minutes and eat as vegetable green.
c. Steam until tender and eat as vegetable green.
d. Add liberally to soups and stews.

SHOOTS:

a. Cut up and put in green salads.
b. Cook shoots (boil or steam), let cool and use in salads.
c. Boil young shoots 10 to 15 minutes or until tender.
d. Steam young shoots until tender.
e. Add shoots generously to soups and salads.

SEEDS:

a. Roast seeds and then chew for nutritious, tasty snack.
b. Add roasted seeds generously to salads.
c. Stir roasted seeds into soups and stews.
d. Grind or crush roasted seeds and use as coffee substitute.

NOTE:

1. Cooled cooked shoots (boiled or steamed) are delicious when eaten with cooked, wild asparagus.
2. Ground and roasted goosegrass seeds far surpass chicory as a coffee substitute (Chapter 2).
3. Older, more mature leaves are tough and must be boiled or steamed to make them edible.

GOURD, WILD (LUFFA SPONGE)

GOURD:

a. Boil whole gourd 15 to 25 minutes or until tender.
b. Steam entire gourd until tender.
c. Roast whole gourd until tender.
d. Bake whole gourd until tender to prick of fork.

LEAVES:

a. Boil young leaves 10 minutes or until tender.
b. Steam young leaves until tender and eat as vegetable.
c. Saute young leaves in lightly greased frying pan (Chapter 2).

FLOWERS:

a. Eat flowers raw for quick snack.
b. Add to green salads.
c. Toss into soups and stews.

SEEDS:

a. Roast seeds over embers or in oven and eat like peanuts.

SHOOTS:

a. Boil shoots 10 minutes or until ready to eat.
b. Steam shoots until tender and ready to eat.
c. Saute in lightly greased frying pan (Chapter 2).

NOTE:

1. Wild gourds grow like cucumbers, watermelons and cantaloupes.
2. Seeds are highly nutritious and improved with salting.
3. Cook gourds (boil or steam) when half ripe for best taste.

GREENBRIER

ENTIRE PLANT IS EDIBLE:

a. Boil or steam until tender and eat as vegetable.
b. Use in soups and stews.

LEAVES:

a. Eat young leaves raw as quick snack.
b. Use uncooked leaves as salad greens.
c. Boil or steam older leaves until tender and eat as spinach.
d. Use liberally in soups and stews.

SHOOTS:

a. Eat young shoots raw as crispy snack.
b. Use crispy young shoots in salads.
c. Boil or steam until tender and eat like asparagus.
d. Cool older shoots after cooking and put in salads.

TENDRILS:

a. Eat tendrils raw as a crispy snack.
b. Use tendrils to liven up salads.
c. Boil or steam with leaves and shoots until tender.

TUBEROUS ROOTS:

a. Wash and crush to obtain jelly-like material. Use in soups and stews as thickener.

NOTE:

1. Drink nourishing liquid used for cooking plant parts.

GROUNDNUT

TUBERS:

a. Peel and eat raw like potatoes for snack.
b. Peel and boil 20 to 25 minutes in heavily salted water as with turnips and potatoes.
c. Wash, peel and roast like potatoes.
d. Wash, peel, slice and fry in bacon grease (if available).
e. Wash and then bake 45 to 60 minutes until tender.
f. Deep fry strips or thin slices of raw tuber like French fries.

SEEDS:

a. Remove from pod and eat bean-like seeds as snack.
b. Remove seeds from pods and add to salads.

c. Remove bean-like seeds from pod and fry until crisp.
d. Boil bean-like seeds until tender and eat.
e. Steam seeds until tender.

PODS:

a. Roast at 375 degrees for 20 to 25 minutes, remove seeds, and fry pod in oil until browned.

NOTE:

1. Tubers are pleasant tasting -- sweet and turnip-like.
2. One of the best wild foods available. Widely used by early settlers and Indians.
3. Tubers are made up of 15% protein.
4. Drink nourishing liquid used for cooking plant parts.

HOG PEANUT

FLOWERS:

a. Eat raw as quick snack.
b. Put raw flowers in green salads.
c. Add liberally to soups and stews.
d. Saute in lightly greased frying pan (Chapter 2).

STEMS:

a. Cut up into small pieces and add to salads.
b. Boil 10 minutes or more and eat as green vegetable.
c. Steam until tender and eat as green vegetable.
d. Put into soups and stews.
e. Saute in lightly greased frying pan (Chapter 2).
f. Cut into small pieces, blend with eggs and fry as omelet.

SEEDS (BEANS):

a. Boil seeds 15 to 20 minutes or until tender.
b. Steam seeds until tender and eat as vegetable.
c. Add liberally to soups and stews.
d. Dry seeds thoroughly and grind or pound into flour.

NOTE:

1. One seed can be found in each pod just below ground level.
2. Drink nourishing liquid used for cooking plant parts.
3. The light brown seeds (beans) from the underground pods are dryer than most.
4. Cooked seeds are good eaten with butter, salt and pepper.

d. Saute in lightly greased frying pan (Chapter 2).
e. Add liberally to soups and stews.

NOTE:

1. Seed pods can be chewed uncooked when fresh.,
2. Drink nourishing liquid used for cooking plant parts.
3. Dry roots thoroughly and grind to make pungent seasoning.

HYSSOP HEDGE-NETTLE

TUBERS:

a. Cut up raw tubers and eat as tasty crispy snack.
b. Cut up raw tubers and add to green tossed salad.
c. Boil tubers until tender and eat as vegetable.
d. Steam tubers 10 to 15 minutes or until tender.
e. Add liberally to soups and stews for flavoring.
f. Saute in lightly greased frying pan (Chapter 2).
g. Use for making delicious fritters (Chapter 2).
h. Cut up and blend with raw eggs to make omelet.

NOTE:

1. Look for crisp white tubers in soil around shriveled stems.
2. Tubers have pronounced bitter flavor. Also tastes minty.

INDIAN CUCUMBER ROOT

TUBERS;

a. Eat crispy tubers uncooked as snack.
b. Cut up crispy tubers and add to salads.
c. Boil tubers as you would potatoes.
d. Steam tubers until done and serve like potatoes.
e. Add tubers to soups and stews.
f. Slice up tubers and fry like potatoes.
g. Saute in lightly greased frying pan (Chapter 2).

NOTE:

1. Tubers have delicate cucumber taste.
2. Tubers are to be gathered during the summer.
3. Berries on this plant are not to be eaten.

LADY'S THUMB (REDLEG)

LEAVES:

a. Put raw chopped leaves in green salads.
b. Boil 10 minutes and eat like spinach.
c. Steam until tender and eat like spinach.
d. Add liberally to soups and stews.

SMALL BULBS (ON LOWER PART OF FLOWER STALKS):

a. Eat raw as tasty snack,
b. Put uncooked bulbs in green salads.
c. Add liberally to soups and stews.

YOUNG ROOTS:

a. Bake until tender and eat like baked potatoes.
b. Roast until thoroughly done and eat like potatoes.
c. Slice raw and fry like potatoes.
d. Cut in strips and fry like French fries.
e. Boil roots 10 to 15 minutes and eat like potatoes.
f. Steam roots until done and eat like potatoes.
g. Saute in lightly greased frying pan (Chapter 2).

NOTE:

1. Serve boiled leaves with vinegar, if available.
2. Roots are tastiest when roasted.
3. Young leaves make decent wild spinach-like dish.
4. Save liquid used for cooking to drink or for making soup.

LAMB'S QUARTER (WILD SPINACH)

ENTIRE PLANT IS EDIBLE:

a. Boil or steam 10 to 15 minutes and eat as vegetable.
b. Crush plant and extract juices to make nutritious juice.
c. Add entire plant to soups and stews.

LEAVES:

a. Eat raw when tender in salads.
b. Boil or steam 10 minutes and eat as you would spinach.
c. Add liberally to soups and stews.
d. Saute in lightly greased frying pan (Chapter 2).
e. Mix with eggs, scramble and fry as delicious omelet.
f. Dry and crush or grind to make flour.

FLOWERS:

a. Add liberally to soups and stews.
b. Use to make delicious fritters (Chapter 2).
c. Blend with fresh eggs and cook in frying pan as omelet.
d. Dry thoroughly and grind or pulverize to make flour.

SHOOTS:

a. Add raw shoots to green salads.

SEEDS:

a. Dry seeds thoroughly and grind or pound to make meal or flour.
b. Boil 10 minutes or until soft to make nourishing cereal.
c. Pulverize seeds to make mush for breakfast.

ROOTS:

a. Dry thoroughly and grind or crush to make flour.
b. Boil or steam until tender and eat as vegetable.

NOTE:

1. Need lots of this plant when cooking as bulk is diminished.
2. Very high in vitamin A and C, calcium, minerals and protein.

3. Save the liquid used for cooking to drink and making soup.
4. One of the best wild plants to eat. Has no harsh taste.
5. Add dark meal made from this plant to regular flour (half and half) when making muffins, bread, biscuits or even pancakes.

LEEK, WILD (RAMP)

LEAVES:

a. Eat raw by themselves as snack.
b. Chop into pieces and add to green salads.
c. Boil leaves in salted water and eat as greens.
d. Steam leaves until tender and eat as greens.
e. Add cut up leaves to soups and stews.
f. Cut into small pieces and blend with eggs to make omelet.
g. Saute in lightly greased frying pan (Chapter 2).

BULBS:

a. Eat raw by themselves as nourishing snack.
b. Chop up and add to tossed salads for taste treat.
c. Boil or steam and eat as vegetable.

d. Saute whole or chopped (Chapter 2).
e. Add whole or chopped to soups and stews.
f. Use to make delicious fritters (Chapter 2).
g. Chop up and blend with eggs to make omelet.

NOTE:

1. Save cooking liquid for drinking or making soups, etc.
2. Wild leeks have a very strong onion smell.
3. Try putting finely chopped bulbs in mashed potatoes.

COOKING POINTERS:

1. Wild leeks are wonderful for stuffing game before cooking.
2. Cook wild leeks with meats and other vegetables.

LETTUCE, WILD

LEAVES;

a. Eat raw by themselves as snack.
b. Use uncooked in salads with other greens.
c. Boil 10 to 15 minutes with 2 water changes.
d. Steam until wilted and serve with butter, salt and pepper.

e. Pour boiling water on leaves. Drain off water after 5 minutes. Sprinkle leaves with fried bacon crumbles for mouth-watering salad.
f. Saute leaves in lightly greased frying pan (Chapter 2).

FLOWERS:

a. Eat raw as snack.
b. Use uncooked in salads.
c. Boil 10 minutes with 2 water changes and eat.
d. Steam until tender and eat as vegetable.
e. Saute in lightly greased frying pan (Chapter 2).
f. Add liberally to soups and stews.
g. Use for making delicious fritters (Chapter 2)

ROOTS:

a. Slice and chew raw as nutritional gum.
b. Slice and boil until tender.
c. Slice and steam until tender.
d. Slice and fry until browned.
e. Slice and saute in lightly greased frying pan (Chapter 2).
f. Cut into small pieces and use to make fritters (Chapter 2).

NOTES:

1. Leaves and flowers when eaten raw are often bitter.
2. Dry roots in sun or over embers and set aside for later use.
3. Eating lots of raw wild lettuce can cause indigestion.
4. Wild lettuce leaves, always slightly bitter, are much better when mixed and eaten with other greens.
5. Save cooking liquid and drink or use for making soup or stew.

COOKING POINTERS:

1. Boiling leaves and flowers in two water changes greatly reduces any bitter taste.
2. To also alleviate bitterness, leaves and flowers can instead be parboiled.
3. Flowerheads added to a casserole give the dish a unique flavor.

LOTUS LILY

LEAVES:

a. Boil unrolling young leaves 10 to 15 minutes.
b. Steam unrolling young leaves until tender.
c. Add unrolling young leaves to soups and salads.

STEMS:

a. Remove rough outer layer and boil until tender.
b. Steam until tender after removing outer layer.
c. Remove rough outer layer and add to soups and stews.

TUBERS:

a. Bake until done as sweet potatoes are baked.
b. Boil until tender as you would sweet potatoes.
c. Steam until tender as you would sweet potatoes.
d. Slice and fry until browned as you would potatoes,

SHOOTS:

a. Boil shoots until tender and eat like spinach.
b. Steam shoots until tender and eat like spinach.
c. Add shoots liberally to soups and stews.

SEEDS:

a. Eat young seeds raw as quick snack.
b. Boil seeds, salt and eat like nuts.
c. Roast seeds until browned and eat like nuts.

d. Roast or bake seeds until dry and pound or grind into flour.
e. Fry as you would popcorn and eat as snack.

ROOTSTOCK:

a. Boil 20 to 25 minutes and eat like sweet potatoes.
b. Mash boiled rootstock and fry like potato patties.
c. Slice raw rootstocks and fry like potatoes.

NOTE:

1. Ripe fruit are nut-like and must be cracked to get seeds.
2. Seeds are edible when ripe after removing bitter embryo.
3. Rootstock may grow to 50" long with many tuberous enlargements.

COOKING POINTERS:

1. Substitute flour or meal made from seeds for part of flour called for in bread, roll and muffin recipes.
2. Peel and mash boiled tubers and add milk, butter and salt. Blend well and serve when smooth like mashed potatoes.
3. Can also fry the above as delicious patties until brown.

MALLOW, COMMON

ENTIRE PLANT IS EDIBLE:

a. Boil 20 minutes or more and eat as vegetable.
b. Steam until tender and eat as vegetable.
c. Use in soups and stews.

LEAVES:

a. Eat raw when young and tender in salads.
b. Boil or steam until tender and eat with butter as a vegetable.
c. Saute in lightly greased frying pan (Chapter 2).
d. Add liberally to soups and stews as thickener like okra.

SHOOTS:

a. Use shoots in salads with raw leaves and fruit.
b. Boil, steam or stew shoots and eat as vegetable.
c. Add liberally to soups and stews.
d. Saute in lightly greased frying pan (Chapter 2).

FRUIT:

a. Eat raw as good snack, or use with leaves and shoots in salads.
c. Boil, steam or stew until tender and ready to eat.

ROOTS:

a. Boil roots to make thick stock for soups.
b. Slice boiled root and fry with chopped onion in lightly greased frying pan until nicely browned.
c. Saute until lightly browned (Chapter 2).
d. Cut into small pieces and use to make fritters (Chapter 2).
e. Cut up root and add liberally to soups and stews.

NOTE:

1. Drink highly nutritional liquid used for cooking plant parts.

COOKING POINTERS:

1. Boil leaves, shoots and roots in water until liquid thickens. Can be used as substitute for egg white in meringue.
2. For delicious candy, cover peeled and sliced roots with water;
   a. Boil 20 to 30 minutes until tender.
   b. Drain off water and add sugar to taste.
   c. Again boil and stir until extremely thick.
   d. Beat mixture and drop spoonfuls on non-stick surface.
   e. When cool, roll each piece in powdered sugar.

MARSH-MARIGOLD (COWSLIP)

YOUNG LEAVES:

a. Boil 10 to 15 minutes or until tender in 3 changes of water.
b. Steam 15 minutes or until tender and eat as vegetable.
c. Use liberally in soups and stews.

FLOWERS:

a. Boil buds 10 to 15 minutes in 3 changes of water.
b. Steam 15 minutes or until tender and eat as vegetable.
c. Stir into soups and stews as nourishing supplement.
d. Use to make tasty fritters (Chapter 2).

STEMS:

a. Boil 10 to 15 minutes in 3 changes of water and eat as greens.
b. Steam 20 minutes or longer and eat as greens.
c. Cook until done and use for making pickles (Chapter 2).
d. Add liberally to soups and stews.
e. Saute in lightly greased frying pan (Chapter 2).

NOTE:

1. Don't eat any of this plant raw. It contains an acrid poison -- poisonous glucoside -- eliminated only by cooking.
2. Do not drink liquid used for cooking these plant parts!
3. Best leaves are obtained before flowers bloom in spring.
4. Few plants will cook up so delightfully in the spring.
5. Eat cooked leaves and stems with butter, salt and pepper.

COOKING POINTERS:

1. This plant, when cooked, requires several changes of water because of its extreme bitterness.

MUSTARD, WILD

LEAVES:

a. Eat chopped pieces of leaves for slightly bitter snack.
b. Eat finely chopped leaves raw in salads.
c. Boil older leaves 30 minutes in 2 water changes.
d. Steam until tender and eat as vegetable green.
e. Use liberally in soups and stews.

FLOWERS:

a. Eat uncooked as snack.
b. Add to salads, soups and stews.
c. Simmer clusters of closed flowers buds 3 to 5 minutes. Do not overcook or they will be ruined!
d. Steam clusters of closed flower buds until done and eat as broccoli-like vegetable.
f. Saute in lightly greased frying pan (Chapter 2).

STEMS:

a. Eat raw as crispy snack.
b. Eat uncooked in salads.
c. Boil until tender and eat as vegetable.
d. Steam for few minutes and eat as vegetable.
e. Add to soups and stews as pungent spice.
f. Saute in lightly greased frying pan (Chapter 2).

SEEDS:

a. Dry thoroughly and use to add zest in soups and stews.
b. Finely grind between 2 stones to make yellow mustard.
c. Add tender fresh seed pods to salads as a great garnish.

NOTE:

1. A wild mustard plant is loaded with vitamins A, Bs and C. It also contains iron, phosphorous, potassium, etc.
2. Flower buds are extremely rich in protein.
3. Do not pick any upper stem leaves when picking flower buds. They are terribly bitter and not edible.
4. Leaves become objectionably bitter as they mature.
5. Green and tender seed pods should be picked only while flowers are in bloom.
6. Save cooking liquid and drink or use to make soup or stew.

COOKING POINTERS:

1. Mustard leaves lose their bulk and shrink when cooked as does spinach and some other greens.
2. Wild mustard requires more cooking time than most other greens.
3. Always cook mustard leaves (greens) at least 30 minutes to cut their pungent biting taste.
4. Flavorful mustard greens should be blended with any of the more tasteless greens such as chickweed or purslane. This will make the bland greens as well as the mustard greens more edible.
5. Boiled flower bud clusters are a great broccoli substitute.

NETTLE, STINGING

ENTIRE PLANT IS EDIBLE:

a. Boil entire plant until tender and eat as vegetable.
b. Steam whole plant until tender and eat as vegetable.
c. Boil entire older plant 30 to 40 minutes, strain out plant, and use cooking liquid for soup stock.
d. Put in soups and stews.

LEAVES:

a. Boil young leaves until tender and eat as green vegetable.
b. Steam young leaves 10 minutes and eat as green vegetable.
c. Add liberally to soups and stews.
d. Make a delicious puree (Chapter 2).
e. Dry and store for later use. Easily reconstituted by steaming or allowing to simmer.

YOUNG SHOOTS:

a. Eat raw for tasty quick snack.
b. Add to green salads.
c. Boil 10 to 15 minutes or less until tender.

d. Steam until tender and eat as green vegetable.
e. Put into pot of boiling water and simmer until tender. Serve with butter, salt and pepper.
f. Add liberally to soups and stews.
g. Saute in lightly greased frying pan (Chapter 2).
h. Stir into eggs and fry as omelet.

NOTE:

1. Best to gather stinging nettle when less then 12" tall.
2. Pick tender green shoots in the spring.
3. Stinging nettle has higher protein content than any other known wild or domestic plant. Also high in vitamin A and C and minerals.
4. One of the tastiest and nutritious plants in the U.S.
5. Always save liquid used for cooking. Add lemon juice and sugar enough to sweeten for refreshing, vitamin loaded drink.

COOKING POINTERS:

1. Liquid used for cooking can also be used as soup by adding salt, pepper and a little vinegar.
2. Cook stinging nettle leaves exactly as spinach is cooked.

3. Stinging qualities of this plant disappear upon cooking.
4. Cook with dandelion greens, wild leeks or wild onions, and wild carrots (Queen Anne's Lace). Eat as vegetable stew.

NUT GRASS

TUBERS:

a. Eat raw as tasty snack.
b. Cut up and put in salads.
c. Boil until tender, peel and eat like small potatoes.
d. Steam in a little water until tender, peel and eat.
e. Bake until soft and eat like baked potato.
f. Bake until done, mash and fry as patties until browned.
g. Slice and fry like potatoes.
h. Cut into strips and deep fry like French fried potatoes.
i. To make delightful coffee substitute:
   1. Roast to dark brown color.
   2. Grind or pound the sweetish, almond flavored tubers to powder.
j. Dry in oven or sun and pound or grind to a powder. Mix half and half with flour for making tasty breads, cakes and cookies.
k. Saute in lightly greased frying pan (Chapter 2).

l. Add liberally to soups and stews.
m. Cut up and use to make delicious fritters.
n. Cut in small pieces, stir into eggs and fry as omelet.

NOTE:

1. To make highly nutritious cold drink:
   a. Cut up tubers and soak in water 3 days.
   b. Take tubers out and crush in fresh water.
   c. Strain and sweeten to taste.
   d. Tubers have nice sweet, nutty taste.
2. Save cooking liquid and drink or use to make soup or stew.

ONION, WILD

ENTIRE PLANT IS EDIBLE:

a. Boil 10 minutes and eat with other vegetables.
b. Steam until tender and eat with other vegetables.
c. Put in soups and stews for flavoring and to eat.
d. Cut into small pieces and eat raw as snack.
e. Cut up and add to salads.
f. Saute in lightly greased frying pan (Chapter 2).

LEAVES:

a. Eat raw as a tasty snack food.
b. Tear up and put in salad with other greens.
c. Boil 3 to 5 minutes and eat as green vegetable.
d. Steam until tender and eat as green vegetable.
e. Add liberally to soups and stews.

FLOWERS:

a. Eat raw by themselves as snack.
b. Put in green salads.
c. Boil or steam and eat as vegetable.
d. Use to make delicious fritters (Chapter 2).

BULB:

a. Eat uncooked for nice tasting snack.
b. Cut up raw and use in green salads.
c. Boil or steam and crush when soft to make onion soup.
d. Boil or steam until tender and eat as vegetable.
e. Boil until tender. Then cream as vegetable dish (see below).

NOTE:

1. Eating lots of wild onions reduces cholesterol.
2. Wild onions taste, look and smell like regular onions.
3. Wild onions are the most common edible wild bulb.

COOKING POINTERS:

1. To make creamed wild onions:
   a. Put 1 tablespoon butter in warm frying pan and melt.
   b. Add dash of pepper, 2 tablespoons flour and 3/4 teaspoon salt. c. Put in 1 cup cut up onion leaves and bulbs.
   c. Stir in milk and let simmer until mixture thickens.
2. To make onion rings:
   a. Slice bulb and dip slices in egg.
   b. Roll in flour or meal.
   c. Drop in pan of hot oil.
   d. Deep fry until nicely browned.

PARSNIP, WILD

LEAVES:

a. Boil until tender in 2 or 3 water changes.
b. Steam until tender and eat as any other green vegetable.
c. Add liberally to soups and stews.

SHOOTS:

a. Boil until tender in 2 or 3 water changes.
b. Steam until tender and eat as any other green vegetable.
c. Add liberally to soups and stews.

ROOTS:

a. Eat cut up pieces of raw root as quick snack, or add to salads.
b. Boil 30 to 45 minutes or until tender.
c. Steam until tender as you would cultivated parsnips.
d. Slice up and fry like potatoes.
e. Bake until soft and eat with butter, salt and pepper.
f. Saute pieces in lightly greased frying pan (Chapter 2).
g. Add liberally to soups and stews.
h. Use for making delicious fritters (Chapter 2)

NOTE:

1. Wild parsnips smell similar to cultivated parsnips.
2. Flavor improves immensely if wild parsnips are not gathered before the first frost.
3. Young leaves and shoots are best if picked in the spring.

COOKING POINTERS:

1. Cook exactly as you would cultivated parsnips.

PENNYCRESS, FIELD

YOUNG LEAVES:

a. Tear up uncooked leaves and use for salads with other greens.
b. Boil in two water changes for 15 to 30 minutes.
c. Steam in two water changes for 15 to 30 minutes.
d. Add liberally to soups and stews.

SHOOTS:

a. Break into small pieces and add to salads with other greens.
b. Boil in 2 water changes for 15 to 30 minutes.
c. Steam in 2 water changes for 15 to 30 minutes.

d. Add shoots liberally to soups and stews.
e. Cut into small pieces, stir into eggs and fry as an omelet.
f. Saute in lightly greased frying pan (Chapter 2).

SEEDPODS:

a. Dry seedpods and use as seasoning somewhat like pepper.
b. Dry seedpods and grind or pound to make flour.

NOTE:

1. Field pennycress is naturally quite bitter, it is classified in the mustard family.
2. Field pennycress is best when eaten like spinach.
3. Serve cooked greens and shoots with butter, salt and pepper.
4. Boiled shoots make an excellent pot herb.
5. Young shoots and leaves should be gathered in the spring.
6. Save cooking liquid and drink or use to make soup or stew.

PEPPERGRASS

LEAVES:

a. Pungent young leaves can be used sparingly in salads with other greens.
b. Boil 10 to 20 minutes with 2 to 3 water changes.

c. Steam until tender with 2 to 3 water changes.
d. Add leaves liberally to soups and stews after first boiling and changing water 2 times.

SHOOTS:

a. Raw young shoots can be sparingly used in salads.
b. Boil shoots 10 minutes. Dump water out. Refill pot. Bring to boil again. Simmer 15 minutes more. Serve as green vegetable.
c. Boil in at least 2 water changes before using as potherb.
d. Add liberally to soups and stews.

SEEDPODS:

a. Use peppery green seedpods sparingly in salads.
b. Add to soups and stews as seasoning.

NOTE:

1. Seeds have a distinctive peppery taste.
2. Peppergrass is another bitter member of the mustard family.3. 3. Peppergrass leaves are loaded with vitamin A and C as well as iron and protein.

4. Leaves and shoots must be boiled with changes of water in order to tone down the high degree of objectional bitterness.

PICKERELWEED

YOUNG LEAVES:

a. Chop up fine and add raw pieces to green salads.
b. Boil or steam until tender and eat as green vegetable.
c. Stir fry in lightly greased pan (Chapter 2).
d. Add liberally to soups and stews.

SEEDS:

a. Eat starch-filled seeds by themselves as snack.
b. Add fresh seeds to green salads.
c. Dry seeds over embers or in sun and add to various cereals.
d. Roast seeds until crisp and brown and pulverize into flour.
e. Add raw or roasted seeds to soups and stews.

PLANTAIN, BROADLEAF

ENTIRE PLANT IS EDIBLE:

a. Soak in salt water 15 minutes and then boil in covered pot 10 to 15 minutes or until tender.
b. Soak in salt water 15 minutes and then steam until tender.
c. Use in soups and stews after soaking 15 minutes in salt water.

LEAVES:

a. Chop up tender young leaves and add to salads.
b. Boil young leaves in covered pot 10 to 15 minutes or until tender and eat like spinach.
c. Soak older leaves in salt water 15 minutes and then boil or steam in covered pot until tender.
d. Saute in lightly greased frying pan (Chapter 2).
e. Use as delicious filler for soups and stews.

SEEDS:

a. Eat raw by themselves as snack or use in salads.
b. Dry seeds over fire or in sun and eat for snack.

c. Roast until browned and eat for nutritious snack treat.
d. Pulverize or grind dry seeds into flour (Chapter 2).

NOTE:

1. Gather young leaves if possible. Older leaves too stringy.
2. Young plantain leaves taste somewhat like mushrooms.
3. Plant is loaded with vitamin A and C as well as many minerals.
4. Save cooking liquid and drink it or use for soups and stews.

COOKING POINTERS:

1. Try not to overcook leaves as they get tough and stringy.

POKEWEED (POKE)

YOUNG LEAVES:

a. Boil 10 to 15 minutes or until tender in 3 water changes.
b. Steam until tender and eat like any other vegetable green.
c. Add liberally to soups and stews.

YOUNG SHOOTS:

a. Boil 15 to 25 minutes in at least 2 changes of water. Lastly add fresh boiling water, cover pot and simmer until tender.

b. Add young shoots liberally to soups and stews.
c. Saute in lightly greased frying pan (Chapter 2).

STEMS:

a. Remove skin and laves from young stem. Throw skin away and boil or steam leaves as greens.
b. Simmer stem until tender and eat as you would asparagus.

NOTE:

1. Large root isn't edible. It is poisonous.
2. Mature leaves aren't edible. They are poisonous.
3. Berries aren't edible. They are poisonous.
4. Seeds aren't edible. They are poisonous.
5. Young shoots with red tinge aren't edible. They are poisonous.
6. Young shoots are loaded with vitamin A.
7. Collect young shoots in early spring before they are 6" to 8" tall or leaves have unfolded.

POTATO VINE, WILD

ROOTS:

a. Peel and boil 20 to 25 minutes until tender.
b. Peel and steam until tender.
c. Bake until soft as you would sweet potatoes.
d. Roast until done as you would sweet potatoes.
e. Peel, slice thin and fry like regular potatoes.
f. Peel, slice in strips and deep fry like French fries.
g. Saute peeled pieces (Chapter 2).
h. Cut up and liberally add to soups and stews.
i. Dry root pieces over embers or in sun. Set aside to use later.

NOTE:

1. Never eat raw roots as they are a purgative (laxative).
2. Some of these roots are extremely bitter.
3. Bitterness can be reduced by changing the water several times while boiling.
4. The large vertical roots are much like yams.
5. Serve potato vine roots with butter, salt and pepper as you would yams.
6. Older roots are much too fibrous or woody to eat.
7. All potato vine roots have a tough skin on the outside.

8. Slice open baked or boiled root and sprinkle with brown sugar while still hot. Then add butter on top of this.
9. Save cooking liquid and drink or use to make soup or stew.

PRICKLY PEAR

FRUIT:

a. Slice top off egg-shaped fruit, peel back skin and eat pulp.
b. Stew fruit pulp and eat as dessert.
c. Boil fruit pulp until it thickens suitably to be syrup.
d. Remove pulp and make jelly or jam (Chapter 2)

STEM:

a. Slice stem lengthwise, cut into strips and eat raw as a snack.
b. Boil or steam strips until soft and eat like string beans.
c. Roast strips over fire, peel and eat.
d. Deep fry strips like onion rings (Chapter 2)

SEEDS:

a. Dry seeds over embers or in sun. Grind or pound to make flour.
b. Use pulverized dry seeds as soup thickener.

NEW PODS:

a. Slice into string bean-like strips and eat raw as snack.
b. Boil or steam until tender and eat as nourishing vegetable.
c. Stir fry or saute in lightly greased frying pan (Chapter 2).
d. Use to make fritters (Chapter 2).
e. Roast over open fire and eat as vegetable dish.
f. Bake until lightly tanned and eat.

NOTE:

1. Remove bristles by rubbing with damp cloth or handful of grass.
2. Emergency water (actually a slightly bitter and sticky juice) can be squeezed or sucked from the stems.
3. Save cooking liquid and drink or use for soup or stew.

PURSLANE

ENTIRE PLANT IS EDIBLE:

a. Cut up and use uncooked in salads.
b. Makes a delicious pot herb for various meat dishes.
c. Boil 10 minutes or until tender and eat as vegetable dish.

d. Steam until tender and eat as delicious vegetable.
e. Add to soups and stews as vegetable and thickener.

LEAVES:
a. Put raw, sharp-tasting leaves sparingly in salads.
b. Boil until tender as a green vegetable.
c. Steam until tender as a green vegetable.
d. Saute until lightly browned (Chapter 2).

SHOOTS:
a. Pick, wash and eat raw as quick snack.
b. Pick, wash and put raw in salads.
c. Boil until tender and eat as green vegetable.
d. Steam until tender and eat as green vegetable.
e. Add liberally to soups and stews.

STEMS:
a. Boil until tender and eat like asparagus.
b. Steam until tender and eat like asparagus.
c. Saute in lightly greased frying pan (Chapter 2).

SEEDS:

a. Dry tiny black seeds in shady place for 7 days. Grind or pulverize for use as flour.

NOTE:

1. Purslane in rather bland. Add chicory, mustard greens or dandelions to make it more edible.
2. The shoots of this plant are one of the best wild shoots when cooked or when used raw in salads.
3. The leaves and stems are rich source of vitamin A and C as well as iron, calcium and phosphorus.
4. To obtain more seeds, simply dry plant and shake in a bag.
5. Save cooking liquid and drink or use to make soup or stew.

COOKING POINTERS:

1. Boil thicker stems 10 minutes and drain. Dry in sun and store in brown paper bag for later use as thickener for soups and stews.
2. Blend ground seeds with equal amount of regular wheat flour for bread-making. The result is a hearty loaf with a unique texture.

QUEEN ANNE'S LACE (WILD CARROTS)

ENTIRE PLANT IS EDIBLE:

a. Excellent in soups and stews as it imparts carrot-like flavor.
b. Chop up and use for making green salad.
c. Boil until tender and eat as green vegetable.
d. Steam until tender and eat as green vegetable.

LEAVES:

a. Eat uncooked in salad.
b. Boil until tender and eat as greens.
c. Steam until tender and eat as greens.
d. Saute leaves until lightly browned (Chapter 2).

FLOWERS:

a. Eat flowers for quick snack.a
b. Eat flowers raw in salads.
c. Saute flowers (Chapter 2)
d. Use for making delicious fritters (Chapter 2).

SEEDS:

a. Dry over embers or in sun and set aside for crunchy snack.

ROOTS:

a. Boil or steam until tender as you would garden carrots.
b. Use as a vegetable in soups and stews.

NOTE:

1. If root doesn't smell like a carrot, don't eat it! It could be dangerous hemlock look-alike.
2. This plant is high in fiber content and vitamin A.
3. Save cooking liquid and drink or use for soup or stew.

REDBUD TREE

FLOWERS:

a. Eat raw as quick snack.
b. Add fresh flowers to green salads.
c. Saute in butter until lightly browned (Chapter 2).
d. Add liberally to soups and stews.

FLOWER BUDS:

a. Eat raw as quick snack.
b. Add to green salads.

c. Boil 5 minutes and eat as vegetable.
d. Steam until tender and eat as vegetable.
e. Saute for 10 minutes in butter (Chapter 2).
f. Make pickled buds (Chapter 2).
g. Add liberally to soups and stews.
h. Use to make delicious fritters (Chapter 2).

YOUNG PODS:

a. Saute flat bean-like pod (Chapter 2).

NOTE:

1. Butter, if available, is better when wanting to saute or stir-fry flowers, flower buds and young pods.
2. Save cooking liquid and drink or use to make soup or stew.

COOKING POINTERS:

1. Flowers, flower buds and young pods can be blended together and sauteed or stir-fried for delicious vegetable dish.

RICE GRAINS:

a. Boiling is simplest method of cooking rice:
   1. The proportions: 2 cups water for each cup rice.
   2. Add 1 teaspoon salt and 1 tablespoon butter.
   3. Bring to boil and stir.
   4. Lower heat, cover pan and simmer 30 to 45 minutes.
   5. Makes about 4 cups (4 to 6 servings) of cooked rice.

b. Eat boiled rice with milk and sugar for snack or cereal.

c. Roast rice grains and grind or pound into fine flour.

d. Blend rice flour with cooking oil to make small cakes:
   1. Add nuts, fruit, meat or anything else to make tasty dish.
   2. Bake or fry cakes until nicely browned.
   3. Wrap in large green leaves and carry for future use as needed.

e. Pop rice by putting in sieve and deep frying until kernels pop.

NOTE:

1. If purplish or pinkish highly poisonous fungi is found among the seeds of rice plants, go elsewhere to gather rice.

COOKING POINTERS:

1. Cooked wild rice is excellent for use as stuffing for animals.
2. Wild rice vastly improves any recipe calling for domestic rice.
3. For extra tender rice add 1/2 cup more water and cook 5 minutes longer.
4. 1 pound uncooked rice (2-2/3 cups) makes 10 cups cooked rice.
5. For baking, substitute ground rice flour for 1/3 of the wheat flour called for in recipe.

ROSE, WILD

LEAVES:

a. Eat raw in salads.
b. Boil until tender and eat as greens.
c. Steam for a few minutes and eat as greens.
d. Saute in lightly greased frying pan (Chapter 2).
e. Add to soups and stews.

FLOWERS:

a. Eat uncooked flowers for snack after snipping off green or white bases.
b. Add fresh, sweet tasting rose petals to salads.

c. Boil flowers and eat as vegetable.
d. Steam flowers and eat as vegetable.
e. Use rose petals to make candy (Chapter 2).
f. Use rose petals to make jelly (Chapter 2).
g. Saute flowers and/or petals (Chapter 2).
h. Add flowers to soups and stews.

ROSE HIPS (FRUITS):

a. Eat raw by themselves for quick and nourishing snack.
b. Put uncooked hips in salads.
c. Jelly or jam can be made with hips and sour apples (Chapter 2).
d. Use hips to make delicious fritters (Chapter 2).
e. Saute hips in lightly greased frying pan (Chapter 2).
f. Add to soups and stews.
g. Dry hips in sun and set aside for later use.
h. Grind or grate dried hips to get powder rich in essential minerals and vitamin C.
i. Sprinkle hip powder on all kinds of cereals.
j. For excellent vitamin supplement: Cover hips with boiling water. Stir in sugar to taste. Set aside for 2 hours. Strain.

SHOOTS

a. Boil tender young shoot and eat as vegetable.
b. Steam tender young shoot and eat as vegetable.
c. Saute in lightly greased frying pan (Chapter 2).
d. Cut into pieces, stir into eggs and fry as omelet.
e. Add to soups and stews.

STEMS:

a. Eat uncooked stems as quick snack.
b. Put raw cut up stems in green salads.
c. Boil until tender and eat as vegetable.
d. Steam until tender and eat as vegetable.
e. Add to soups and stews.

NOTE:

1. Peel stems before eating raw or cooking as vegetable.
2. Rose hips have a delicate apple-like flavor.
3. Save cooking liquid and drink or use for soups and stews.
4. Dark roses taste stronger. Lighter flowers more subdued.
5. Rose hips are outstanding survival food. Excellent for preventing scurvy.

6. Rose hips are rich in vitamin C, much higher than oranges.
7. To use rose hips, just cut away the mature orange-red fruit. Remove seeds and eat raw.
8. Seeds are rich in vitamin E.

SHEEP SORREL (COMMON SORREL)

LEAVES:

a. Eat raw for thirst quenching snack.
b. Add raw leaves to salads and liven them up.
c. Boil or steam 3 to 5 minutes in 2 water changes until tender.
d. Add liberally to soups and stews.
e. Saute with eggs and butter (Chapter 2).
f. Steep 30 minutes or longer in boiling water. Strain. Sweeten to taste. Chill and drink as nutritional juice.

FLOWERS:

a. Eat raw for quick snack.
b. Put uncooked flowers in salads.

STEMS:

a. Nibble raw stems for quick snack.
b. Add to salads to liven them up.
c. Boil until tender and eat as vegetable.
d. Steam a few minutes and eat as vegetable.
e. Saute in lightly greased frying pan (Chapter 2).

SEEDS:

a. Add sour tasting seeds to salads.

NOTE:

1. Sheep sorrel has a mildly sour vinegary taste.
2. Eat in small quantities only. Contains oxalic acid.
3. The absorption of calcium may be blocked if large amounts of sorrel is eaten over a long period of time.
4. This plant is loaded with vitamin C.

SHEPHERD'S PURSE

ENTIRE PLANT IS EDIBLE:

a. Cut up plant and eat raw for snack or use in salads.
b. Boil until tender in 2 water changes and eat as vegetable.
c. Steam plant until tender and eat as vegetable.
d. Use entire plant in soups and stews.

LEAVES:

a. Eat raw leaves by themselves for snack or use to make salad.
b. Boil dandelion-like leaves until tender in 2 changes of water and eat as highly nutritious green vegetable.
c. Steam leaves until tender and eat as vegetable greens.
d. Add leaves liberally to soups and stews.
e. Dry leaves and pulverize to use as flour.

FLOWERS:

a. Eat raw as quick snack.
b. Add uncooked flowers to salads.
c. Saute until lightly browned (Chapter 2).

SEED PODS:

a. Boil heart-shaped seed pods until tender.
b. Steam heart-shaped seed pods until tender.
c. Dry seed pods and pulverize to use as flour.
d. Dry seed pods and grind to make meal for cereal.
e. Dry seed pods and use for pepper-like seasoning in soups.

NOTE:

1. Some leaves may be too peppery tasting for salads.
2. Save cooking liquid and drink or use for soups and stews.

SOLOMON'S SEAL

YOUNG SHOOTS:

a. Chop uncooked shoots into small pieces and blend with other leafy plants in salads
b. Boil 15 to 20 minutes and eat as you would asparagus.
c. Steam until tender and eat as you would asparagus.
d. Saute in lightly greased frying pan (Chapter 2).
e. Cut in small pieces, stir into eggs and fry as omelet.

ROOTS:

a. Boil 20 to 30 minutes and eat like potato.
b. Steam until tender and eat like potato.
c. Bake 30 minutes or until tender and eat like potato.
d. Roast on embers or in oven until tender and eat like potato.
e. Cut up and add to soups and stews.
f. Slice and fry until lightly browned like potatoes.
g. Cut up and saute in lightly greased frying pan (Chapter 2).

NOTE:

1. Best to gather roots in fall.
2. Baked, boiled or roasted, these roots taste much like parsnips.
3. Rootstocks are loaded with starch.
4. Save cooking liquid and drink or use to make soup or stew.

COOKING POINTERS:

1. Remove leafy heads from young shoots before cooking. The leafy heads become bitter when cooked.

SPRING BEAUTY

LEAVES:

a. Eat crispy leaves as vitamin-rich snack.
b. Chop leaves and add to green salads.
c. Boil 3 to 5 minutes in salted water and eat as greens.
d. Steam until tender and eat as any other green vegetable.

CORMS (ENLARGED STEM BASE):

a. Boil until fork punctures tough outer skin. Peel off skin. Eat with butter, salt and pepper like potatoes.
b. Boil 15 minutes. Peel tough outer skin. Mash like potatoes.
c. Bake until soft, peel off tough outer skin and eat with butter.
d. Peel off tough outer skin, slice and fry until browned.
e. Saute slices of corm in lightly greased frying pan (Chapter 2).

TUBERS:

a. Eat tubers raw as starchy snack or cut up and add to salads.
b. Boil 10 to 20 minutes in salted water with jacket intact. Then peel off jacket and eat whole or mash as you would a potato.

c. Boil in salted water until tender, slice thin and fry.
d. Bake tubers until soft, peel and eat like baked potato.

NOTE:

1. Leaves are edible but not outstanding when cooked.
2. Tubers and corms taste like new potatoes and cooked chestnuts.
3. A corm is nothing more than an enlarged bulb-like stem base. It's solid rather than in layers like an onion.

SPINY-LEAVED SOWTHISTLE

YOUNG LEAVES:

a. Cut off spines. Tear up leaves if not extremely bitter. Mix with other more bland greens in salad.
b. Cut off spines. Boil leaves 5 minutes in very little water. Change water and boil 15 minutes or until tender.
c. Cut off spines. Steam until tender. Eat as spinach.
d. Add liberally to soups and stews.
e. Cut off spines, chop up leaves, blend with eggs and fry for outstanding omelet.

FLOWERS:

a. Eat by themselves after twisting green base from petals.
b. Eat raw in tossed salads with leaves, shoots and roots.
c. Saute flowers and buds (Chapter 2).
d. Use to make terrific fritters (Chapter 2).
e. Crush dried flowers and add to breakfast cereals.

NOTE:

1. Flower petals have pleasantly sweet taste.
2. Cooked leaves are good served with butter or vinegar.
3. Many people enjoy bitter taste of this plant when cooked.
4. Save cooking liquid and drink or use to make soup or stew.

COOKING POINTERS:

1. If leaves are too bitter, simply mix with lamb's quarter or other more bland greens such as purslane or chickweed.

SUNFLOWER

SEEDS:

a. Eat raw, unshelled seeds as a highly nutritious quick snack.
b. Dry seeds over embers or in sun, or eat inside of roasted nut for tasty snack.
c. Eat boiled seeds for nutritious snack.
d. Boil crushed seeds to extract the pale yellow vegetable oil. Simply skim oil from surface of water.
e. Roast and pulverize or grind into flour.
f. Roast and pulverize or grind shells to make coffee (Chapter 2).
g. Extract nut meats from seed shells. Grind to paste. Add maple syrup or honey to make great peanut butter substitute.
h. Eat raw, roasted or dried seeds as cereal or with cereal.

FLOWERS:

a. Use to make delicious fritters (Chapter 2).

NOTE:

1. Sunflower seeds are high in the B vitamins.
2. Parched seeds can be pound or ground between 2 stones as Indians used to do in making bread.

COOKING POINTERS:

1. Roasted nuts from sunflower seeds can be used in any recipe calling for nuts of any kind.
2. Substitute part of flour called for in bread, roll and muffin recipes with flour made from sunflower nutmeats.

SWEET FLAG (CALAMUS)

YOUNG SHOOTS:

a. Eat raw as pleasant smelling and tasty snack.
b. Use uncooked as excellent addition to salads.
c. Boil 3 to 5 minutes and eat as green vegetable.
d. Steam until tender and eat as green vegetable.
e. Add liberally to soups and stews.

ROOTSTOCK:

a. Use to make delicious candy.
   1. Peel rootstock and cut into 1" lengths.
   2. Boil 1 to 2 hours until tender. Change water 4 or 5 times.
   3. Now make the candy (Chapter 2).
b. Cut into pieces and add to soups and stews.

NOTE:

1. Leaves and rootstock have pleasant spicy smell.
2. Inner part of young stalks are packed with sweet half-formed leaves. That's why they're so great as snacks and in salads.
3. Dried roots can be pounded to powder and used as a natural insecticide.
4. Save cooking liquid and drink or use to make soup or stew.
5. Be extremely careful not to confuse sweet flag with blue flag.
   a. Blue flag is poisonous. Sweet flag is not.
   b. Blue flag leaves are a dull, bluish-green and have no smell.
   c. Sweet flag leaves are glossy, yellow-green with a spicy smell.

THISTLE, BULL

ENTIRE PLANT IS EDIBLE:

a. Boil until tender and eat as vegetable.
b. Steam until tender and eat as vegetable.
c. Add to soups and stews.

LEAVES:

a. Eat raw as snack after cutting away barbed spines.
b. Add to green salads after stripping away barbed spines.

c. Boil until tender after cutting away barbed spines,
d. Steam until tender after cutting away barbed spines.

YOUNG STEMS:

a. Peel and eat raw as snack. They are crispy, crunchy and exceptionally tasty.
b. Peel and add to salads.
c. Boil peeled stems until tender and eat as vegetable.
d. Steam peeled stems until tender and eat as vegetable.
e. Stir into soups and stews.
f. Fry until nicely browned as you would potatoes.
g. Saute stems (Chapter 2).

EARLY SHOOTS:

a. Eat as tasty snack.
b. Add to green salads.
c. Boil until tender and serve as green vegetable.
d. Steam until tender and serve as green vegetable.
e. Fry until browned as you would potatoes.
f. Saute in lightly greased frying pan (Chapter 2).

ROOT:

a. Peel, boil until soft and eat as potato.
b. Peel, steam until soft and eat as potato.
c. Peel, slice and fry until nicely browned.
d. Roast until brown, let cool and pound or grind into flour.
e. Boil until roots turn to mush. Set aside to dry. Then grind or pound into flour.
f. Stir root pieces into soups and stews.
g. Saute pieces in lightly greased frying pan (Chapter 2).

NOTE:

1. Save cooking liquid parts and use for soups and stews.
2. Most nourishing part of the thistle is flower stem when peeled.
3. Leaves , stems and shoots are somewhat like celery.
4. The roots of new thistles are one of best wild survival foods.

VIOLET, SWEET

ENTIRE PLANT IS EDIBLE:

a. Boil 2 to 3 minutes and eat as green vegetable.
b. Steam 2 to 3 minutes and eat as green vegetable.
c. Use liberally in soups and stews.

LEAVES:

a. Eat raw as snack.
b. Eat raw with flowers and buds in green salads.
c. Boil in covered pot 3 to 5 minutes and eat as vegetable.
d. Steam in covered pot until tender and eat as green vegetable.
e. Add to soups and stews as an okra-like thickener.
f. Saute in lightly greased frying pan (Chapter 2).
g. Use to make jelly and jam (Chapter 2).

FLOWERS:

a. Pick blossoms from plant and eat raw as tasty snack.
b. Eat raw in salads with raw young leaves.
c. Boil 2 to 3 minutes and eat with leaves or other greens.
d. Steam until soft and eat with leaves or other greens.

e. Use liberally in soups and stews.
f. Saute in lightly greased frying pan (Chapter 2).
g. Use to make candy (Chapter 2).

ROOTSTOCK:

a. Boil until tender and eat as vegetable.
b. Steam until soft and eat as vegetable.
c. Roast until done and eat as vegetable.
d. Bake until soft and eat as vegetable.
e. Slice and fry until browned.
f. Slice and saute in lightly greased frying pan (Chapter 2).
g. Cut in pieces and use to make fritters (Chapter 2).
h. Use liberally in soups and stews.

NOTE:

1. Violets are extremely high in vitamin A and C.
2. Save nourishing liquid used for cooking plant parts and drink or use for soups and stews.
3. Violet leaves are quite bland and are best when combined with mustard, dandelion or other stronger tasting greens.
4. Violet greens are eaten much like spinach.

COOKING POINTERS:

1. Add flowers to flour when baking for unusual flavor.
2. Barely cover plant parts when boiling.

WAPPATO (DUCK POTATOES)

TUBERS:

a. Eat raw as snack although mildly bitter.
b. Boil or steam until tender and eat like potatoes.
c. Boil until tender, peel, mash and eat as mashed potatoes.
d. Boil until tender, peel and mash. Make patties and fry like hash browns.
e. Roast until lightly browned and eat as potato.
f. Saute in lightly greased frying pan (Chapter 2).
g. Bake like potatoes, peel when done and eat.
h. Slice and fry like potatoes.
i. Dry tuber and grind or pound into flour.
j. Cut up and make a creamed or scalloped potato-like dish.
k. Cut into strips and deep fry like French fried potatoes.

NOTE:

1. Tubers form in the fall.
2. Tubers have mildly unpleasant taste when eaten raw.
3. To get tubers, dig down in mud around plant and cup hands around tubers. Pull slowly up and out.
4. Tubers are starchy and highly nutritious.
5. To dry tubers for storage:
   a. Boil 30 minutes, drain well and cut into 1/4" to 1/2" slices.
   b. Lay in warm oven or the sun to thoroughly dry.
6. To use at later date:
   a. Soak in water 15 minutes.
   b. Bring to boil and let simmer 20 to 30 minutes.

WATER CHESTNUT

FLOATING LEAVES:

a. Tear into small pieces and use in salads.
b. Boil or steam 10 minutes and eat as green vegetable.
c. Add liberally to soups and stews.

SEEDS (NUTS AND MEAT):

a. Boil 10 to 15 minutes, eat as quick snack or add to salads.
b. Steam until tender and eat as snack or add to salads.
c. Roast until crunchy and eat as quick snack.
d. Roast and grind or pound into flour substitute.

NOTE:

1. Water chestnut is extremely high in nutrients.
2. Save cooking liquid and drink or use in soups and stews.

WATERCRESS

LEAVES:

a. Eat raw as a quick snack or add to salads.
b. Boil or steam until tender and eat like spinach.
c. Add liberally to soups and stews.
d. Saute in lightly greased frying pan (Chapter 2).
e. Tear into pieces, add to eggs and fry as omelet.

STEMS:

a. Cut up and eat raw as snack or add to salads as garnish.

b. Boil or steam until tender and eat as vegetable.
c. Add liberally to soups and stews for flavor.

BLOSSOMS:

a. Eat raw as delicious snack or add to salads as garnish.
b. Make complete salad with raw blossoms, leaves and stems.
c. Add to soups and stews.
d. Put peppery flavored blossoms on sandwiches for taste treat.

NOTE:

1. Only plant parts growing above water are edible.
2. Watercress, rich in vitamin C and minerals, is one of most desirable wild greens known.
3. Most people prefer to eat watercress raw than cooked.

COOKING POINTERS:

1. Stuff fish with watercress (after cleaning). Then smoke, fry or bake. Add garlic if available.
2. Makes good soup by itself, especially when potatoes are added.
3. Adds flavor when mixed with bland tasting greens.

WATER LILY, FRAGRANT

LEAVES:

a. Eat young unfurled raw leaves as snack.
b. Add young unfurled raw leaves to green salads.
c. Boil young unfurled leaves 5 to 10 minutes and eat.
d. Steam young unfurled leaves until tender and eat as vegetable.
e. Add liberally to soups and stews.

FLOWERS:

a. Eat white flower petals by themselves as excellent snack.
b. Eat unopened raw buds for delicious snack or use in salads.
c. Use white petals in salads with all other flower parts.
d. Boil unopened buds 5 to 10 minutes and eat with butter.
e. Steam unopened buds until tender and eat with butter.
f. Saute buds in lightly greased frying pan (Chapter 2).
g. Use for making delicious fritters (Chapter 2).

TUBERS:

a. Peel, slice and then boil tubers until tender.
b. Mash boiled tubers and eat like mashed potatoes.

c. Dry mashed tubers and use as flour for baking.
d. Bake like small potato until soft.
e. Cut sweetish tuber in pieces and add to soups and stews.
f. Slice thin and fry like potatoes.
g. Roast until nicely browned.

SEEDS:

a. Fry and pop large seeds as you would popcorn and eat.
b. Seeds make excellent breakfast with milk and sugar.
c. Parch (dry by heat) seed kernels and grind or pound. Use as delicious substitute for cornmeal creamed-corn.
d. Crack seed shells and cook nutmeats as you would rice.

NOTE:

1. Tubers are about the size of a hen's egg.
2. Seeds are high in protein, oil and starch.
3. Tubers contain the most starch from autumn to Early spring.
4. Roots peel easily after boiling, steaming or roasting.
5. Tubers must be loosened from the mud with your feet. Gather them up as they float to the surface.
6. Save cooking liquid and drink or use to make soup or stew.

COOKING POINTERS:

1. If tuber flavor is objectionable, boil in 3 changes of water.

WATER PLANTAIN

LEAVES:

a. Eat by themselves as snack or add to salads.
b. Soak in salt water 15 minutes and boil in covered pot.
c. Soak in salt water 15 minutes and steam until tender.
d. Add liberally to soups and stews.

ROOTSTOCKS:

a. Boil until tender and eat like potatoes.
b. Steam until tender and eat like potatoes.
c. Bake until soft and eat like baked potato.
d. Cut up and add liberally to soups and stews.
e. Slice and fry like potatoes.
f. Mash boiled rootstock, make patties and fry until crisp.

SEEDS:

a. Dry in sun and grind or pound into flour.

e. Add liberally to soups and stews.
f. Stir fry in 2 or 3 tablespoons butter and water.

SEEDS:

a. Roast lightly and eat as nutritional snack.
b. Hull and grind or pound into palatable flour. Do this when seeds appear tan rather than green.

ROOTS:

a. Boil until tender as you would carrots.
b. Steam until tender as you would carrots.
c. Roast until done as you would carrots.
d. Cut up and add liberally to soups and stews.

NOTE:

1. Boiled or steamed young leaves taste like beet greens.
2. Extremely high in vitamin A, minerals and vegetable protein.
3. Save cooking liquid and drink or use for soups and stews.
4. Curly dock loses little of its bulk when boiled or steamed.

COOKING POINTERS:

1. If leaves are bitter when boiled, use 2 to 3 water changes to eliminate bitterness and make them tender.
2. Use seed flour by blending half and half with regular flour.

5

FERNS AS A GOOD FOOD SOURCE

Ferns are abundant in moist areas of all climates. They are especially easy to find. Look in gullies, on stream banks, in forested areas, along the sides of hiking trails and on the edge of woods.

Ferns, by and large, are a safe plant to cook and eat. Some are distastefully bitter and certainly not palatable. Yet, no fern is known to be poisonous.

BRACKEN FERN

STALK OR STEM:

a. Eat fresh stems raw in small quantities as snack.
b. Cut up raw stems and add small amounts to a green salad.
c. Boil stems and eat as you would asparagus.
d. Steam and eat as vegetable green.
e. Add to soups and stews.

f. Cut in pieces, stir into eggs and fry as an omelet.
g. Saute stalks-stems (Chapter 2).
h. Deep fry cut up pieces (Chapter 2).

FIDDLEHEADS (YOUNG SHOOTS-LEAVES BEFORE THEY UNFURL):

a. Eat fresh shoots raw in small quantities as snack.
b. Eat small amounts in green salad.
c. Boil and eat as vegetable.
d. Steam and eat as vegetable.
e. Add to soups and stews.
f. Stir into eggs and fry as delicious omelet.
g. Saute fiddleheads (Chapter 2).
h. Deep fry fiddleheads (Chapter 2).
i. Use for making fritters (Chapter 2).

NOTE:

1. It's best to not eat this fern raw in large quantities on a regular basis. Cooked fern is always fine to eat.
2. Gather top 4" to 6" of young fern leaf. Break off and toss aside the curled wooly covered tips.
3. Drink nourishing liquid used to cook plant parts.

EDIBLE FERN PARTS IN GENERAL

STALK OR STEM:

a. Eat raw as snack or cut up and add to green salad.
b. Boil in salted water and eat like asparagus.
c. Steam and eat as green vegetable.
d. Add to soups and stews.
e. Cut in pieces, stir into eggs and fry as an omelet.
f. Saute stalks-stems (Chapter 2).
g. Deep fry cut up pieces (Chapter 2).

FIDDLEHEADS (YOUNG SHOOTS-LEAVES BEFORE THEY UNFURL):

a. Eat fresh shoots-leaves raw as snack.
b. Eat fiddleheads in a green salad.
c. Boil or steam and eat as vegetable.
d. Add to soups and stews.
e. Stir into eggs and fry as delicious omelet.
f. Saute fiddleheads (Chapter 2).
g. Deep fry fiddleheads (Chapter 2).
h. Use for making fritters (Chapter 2).

NOTE:

1. Break off stalk and draw closed hand over it. This removes the inedible fuzz or wool-like material.
2. Select young stalks during April and May for best eating. Break off as long as they remain tender.
3. Drink nourishing liquid used to cook plant parts.

OSTRICH FERN

STALK OR STEM:

a. Eat fresh stems raw as snack.
b. Cut up raw stems and add to a green salad.
c. Boil or steam until tender and eat like asparagus.
d. Add to soups and stews.
e. Cut in pieces, stir into eggs and fry as an omelet.
f. Saute stalks (Chapter 2).
g. Deep fry cut up pieces (Chapter 2).

FIDDLEHEADS (YOUNG SHOOTS-LEAVES BEFORE THEY UNFURL):

a. Eat fresh shoots raw as snack.
b. Eat raw in a green salad.

c. Boil and eat as vegetable.
d. Steam and eat as a vegetable.
e. Add to soups and stews.
f. Stir into eggs and fry as delicious omelet.
g. Saute fiddleheads (Chapter 2).
h. Deep fry fiddleheads (Chapter 2).
    1. Use for making fritters (Chapter 2).

NOTE:

1. Collect fiddleheads (coiled shoot) during April and May while under 6" high.
2. Wash or scrape off inedible brown scales off in cold water.
3. Drink nourishing liquid used to cook plant parts.

6

TREES -- AN EXCELLENT FOOD SUPPLY

The inner bark of some trees -- the layer next to the wood -- can be eaten raw or cooked. Avoid the outer bark. It contains large amounts of tannin and is extremely bitter.

Flour can be made by pulverizing the inner bark of a number of trees – aspen, birch, cottonwood, pine, slippery elm and willow.

One outstanding example of a tree food source is the pine.

The inner bark is high in vitamin C. The nuts, needles, twigs and sap are all edible. The nuts (eaten raw or roasted) grow in woody cones hanging near the tips of the branches. When mature, they fall out of the ripe cone.

SYRUP TREES

Certain kinds of trees are known to provide nourishing sap. They are invaluable to a person stranded in the wilderness who is desperately trying to stay alive. The sugar maple is the best known and most widely used source of sap for making sweet syrup and sugar. Others used for this are the birch, butternut, hickory and sycamore. Each has to be tapped and the sap collected in buckets or cans. The sap is then boiled down to syrup and further down if sugar (Chapter 2) is the desired end result. The sap from the above mentioned trees produce a syrup comparable to maple syrup.

These trees are also excellent sources of drinking water in a survival situation. This is especially important to know if you happen to be in an area where the water is polluted or in short supply.

BIRCH TREE

INNER BARK:

a. Eat raw strips as good emergency rations.
b. Dry and pound or grind into flour for making bread.
c. Cut in strips and boil as noodles in soup or stew.

TWIGS:

a. Dry thoroughly and eat as crispy treat.

BUDS:

a. Eat uncooked as quick snack or add raw to green salads.
b. Cook (boil or steam) and eat as vegetable.
c. Add liberally to soups and stews.
d. Saute buds (Chapter 2).
e. Use for making fritters (Chapter 2).

LEAVES:

a. Cook (boil or steam) until tender.
b. Saute leaves (Chapter 2).

SAP:

a. Drink nutritious sap in spring as it comes from tree.
b. Boil sap down to make delicious syrup and sugar (Chapter 2).

NOTE:

1. Drink nourishing liquid used to cook tree parts.
2. The sap from every kind of birch tree is edible.
3. Sap flow is copious but only half as sweet as maple.
4. Dry inner bark in sun and put in sealed jars.

BUTTERNUT TREE (WALNUT FAMILY)

NUTS:

a. Eat raw as snack or add to salad made of various greens.
b. Use nuts in baking -- cakes, pies, breads, etc.
c. Roast nuts on embers of fire or in oven.
d. Add nut meats to soups and stews.
e. Mash boiled nuts like potatoes when ready to eat.
f. Boil crushed nuts, skim oil from surface of water and use as cooking oil.
g. Cool the oil and use as excellent butter substitute.
h. Grind or crush nuts to make peanut butter substitute.
i. Use nut meats to make candy (Chapter 2).
j. Pulverize or grind roasted nuts to make excellent coffee substitute (Chapter 2).
k. Grind or pulverize dried nuts and eat as grits or gruel.

SAP:

a. Drink uncooked sap in spring as it comes from tree.
b. Boil sap down to make delicious syrup and sugar (Chapter 2).

NOTE:

1. Nuts and sap are both top notch survival foods.
2. Gather nuts in autumn as they fall from trees.
3. Store nuts in their own shells whenever possible.
4. Nut meats get rancid and inedible if exposed to moisture, heat and light.

HICKORY

NUTS:

a. Eat hickory nuts raw as nutritious snack.
b. Add nuts to salad made up of various greens.
c. Use nuts in baking -- cakes, pies, breads, etc.
d. Roast nuts on embers of fire or in oven.
e. Boil the kernel that lies within the shell.
f. Mash boiled nuts like potatoes when ready to eat.
g. Boil crushed nuts and skim oil from surface of water.
   1. Hickory nut oil makes a great cooking oil.
   2. Cool the oil and use as a butter substitute.
h. Grind or crush nuts to make peanut butter substitute.
i. Use nut meats to make candy (Chapter 2)

SAP:

a. Drink uncooked sap as it comes from tree.
b. Boil sap down to make delicious syrup (Chapter 2).
c. Boil down further to make sugar (Chapter 2).

NOTE:

1. Nuts and sap are both top notch survival foods.
2. Hickory trees are to be tapped for sap in late winter.
3. Gather nuts in autumn as they fall from trees.
4. Separate some nuts from their shells by boiling.
5. Boiled hickory nuts can be ground and mixed with cornmeal, potatoes or sweet potatoes for a tasty dish.

MAPLE TREE

INNER BARK:

a. Eat raw strips as good emergency food.
b. Dry and pound into flour for making bread.
c. Cut in strips and boil as noodles in soup or stew.
d. Cook (boil or steam) until tender and eat as vegetable.

TWIGS:

a. Dry in sun and eat as crispy treat.

SEEDS:

a. Eat raw, roasted, boiled or dry seeds as quick snack.

YOUNG LEAVES:

a. Eat raw by themselves for snack or use as salad green.
b. Boil (or steam) until tender and eat as vegetable.
c. Saute leaves (Chapter 2).

SAP:

a. Drink nutritious sap in spring as it comes from tree.
b. Boil sap down to make delicious syrup (Chapter 2).
c. Boil down further to make maple sugar (Chapter 2).

NOTE:

1. Maple syrup is one of the best kinds of survival food.
2. Maple trees should be tapped for sap in early spring.
3. Best sap flow requires cold nights followed by warm days.
4. About 35 gallons of sap makes one gallon of maple syrup.
5. Leaves are very rich in sugar content.

SYCAMORE

INNER BARK:

a. Eat raw strips as good emergency rations.
b. Dry and pound or grind into flour for making bread.
c. Cut in strips and boil as noodles in soup or stew.
d. Boil or stream until tender and eat as vegetable.

TWIGS:

a. A crispy treat when dried and then eaten as snack.

SEEDS:

a. Chew raw or roasted seeds as quick snack when hungry.
b. Dry seeds in sun or over embers and eat for tasty snack.

YOUNG LEAVES:

a. Eat raw by themselves for quick snack.
b. Tear in pieces and use as salad green.
c. Cook (boil or steam) and eat as vegetable (Chapter 2).
d. Saute leaves (Chapter 2).

SAP:

a. Drink nutritious sap in spring as it comes from tree.
b. Boil sap down to make delicious syrup or sugar (Chapter 2).

NOTE:

1. Sycamore sap is one of the best kinds of survival foods.
2. Save cooking liquid and drink or use for soups and stews.
3. Sycamore sap is excellent substitute for drinking and cooking water when no unpolluted water is available.

MORE TREES WITH EDIBLE NUTS

Nuts can be eaten either fresh or dried, cooked or uncooked. They contain much valuable protein. Great food value is derived from a nut's oil content. A single pound of nut meats provides a person with around 3,000 calories. Nuts are unquestionably among the most nutritious of all plant foods. They are extremely useful in a survival situation.

AMERICAN BEECH

NUTS:

a. Eat beechnut kernels raw as nutritious snack.

b. Slowly roast nuts and eat them whole for snack.
c. Add nuts to salad made up of various greens.
d. Use nuts in baking -- cakes, pies, breads, etc.
e. Add nut meats to soups and stews.
f. Boil crushed nuts and skim oil from surface of water to make cooking oil and butter substitute.
g. Grind or crush nuts to make peanut butter substitute.
h. Pulverize roasted nuts to make coffee substitute (Chapter 2).
i. Use nuts to make candy (Chapter 2).

INNER BARK:

a. Eat raw strips as good emergency rations.
b. Dry and pound or grind into flour for making bread.
c. Cut in strips and boil as noodles in soup or stew.

SAWDUST:

a. Boil or roast sawdust and blend with flour for making bread.

LEAVES:

a. Cook (boil or steam) young leaves in the spring.

NOTE:

1. Thin shelled beechnuts are small, sweet and delicious.
2. Heat nuts and briskly rub together to remove shells.
3. Gather nuts just after they have fallen in October.

CHESTNUT

NUTS:

a. Eat chestnuts raw as nutritious snack.
b. Add nuts to salad made up of various greens.
c. Roast nuts on embers of fire or in oven and eat.
d. Boil crushed nuts and skim oil from surface of water.
    1. Chestnut oil makes a great cooking oil.
    2. Let oil cool and use as excellent butter substitute.
e. Grind or crush nuts to make peanut butter substitute.
f. Grind or pulverize dried nuts and eat as grits or gruel.
    1. Grind dried nuts finer and blend with equal amount of cornmeal.
    2. Shape into small cakes or cookies and bake until crispy.
g. Mash boiled nuts and eat as you would mashed potatoes.
h. Use nut meats to make candy (Chapter 2).

i. Add nut meats to soups and stews.
j. Put cooked or uncooked nutmeats in various desserts.
k. Grind roasted nuts to make coffee substitute (Chapter 2).

NOTE:

1. Pick nuts off ground in late summer and early fall.
2. Both raw and cooked nuts are one of best survival foods.

COOKING POINTERS:

1. Blend acorn nutmeat meal with regular baking flour. Adds agreeable nutty flavor to the cooked or baked items.
2. Put nutmeats with cornmeal to make terrific stuffing.

WHITE OAK

NUTS:

a. Eat acorns raw or cooked for nutritious snack.
b. Add acorn nuts to salads made up of various greens.
c. Roast acorns on embers of fire or in oven and eat.
d. Boil crushed nuts and skim oil from surface of water to make cooking oil and butter substitute.

e. Crush acorn nuts to make peanut butter substitute.
f. Pulverize dried nuts and eat as oatmeal-like cereal.
    1. Grind nuts finer and blend with equal amount of cornmeal.
    2. Form into small cakes and cookies and bake until crispy.
g. Use acorn nut meats to make candy (Chapter 2).
h. Add acorn nut meats to soups and stews.
i. Grind roasted nuts to make coffee substitute (Chapter 2).
j. Mash boiled nuts and eat as you would mashed potatoes.
    1. Make into patties and fry as you would potato patties.

NOTE:

1. Most acorns aren't edible because of bitterness caused by an excess of tannin.
2. Only the white oak family of trees have sweet acorns.
3. Put acorn nuts with cornmeal to make stuffing for game.
4. Add acorn meal to baking flour for nutty flavored bread.
5. Leave acorns in their shells to store.
6. Pick acorns off ground in late summer and early fall.
7. Raw and cooked acorns are one of best survival foods.

7

MUSHROOMS TO EAT OR NOT TO EAT

There are many misconceptions about mushrooms and their imagined food value. In reality, they offer absolutely **nothing** in the way of nutrition. And they're terribly difficult to digest. It's hardly worth the effort required to bend over and pick one up! It's certainly ridiculous to waste time and energy hunting for, cooking and eating mushrooms.

Even the world's most widely respected survival experts can make mistakes regarding mushrooms. John "Lofty" Wiseman, for example, makes these erroneous statements in his otherwise excellent video presentation: OUTDOOR SURVIVAL: SAS SURVIVAL TECHNIQUES:

1. "Fungi [mushrooms]: slightly more nutritional value than plants."
2. "Fungi [mushrooms] come above plants on the nutritional ladder. Pound for pound they got more nourishment."

This is simply not the case and John certainly should have known better!

Then why even bother learning to identify edible mushrooms? Why go to the trouble of preparing mushrooms to eat? And better yet, why eat mushrooms in the first place? No logical answers can be given to the above questions.

On the other hand, the U.S. military takes another approach to the identifying, collecting and eating of mushrooms. The official Army survival manual, FM 21-76, gives this warning: "**Do not eat mushrooms in a survival situation! The only way to tell if a mushroom is edible is by positive identification. There is no room for experimentation. Symptoms of the most dangerous mushrooms affecting the central nervous system may show up after several days have passed when it is too late to reverse their effects.**"

Need more be said?

8

THE ART OF MAKING NOURISHING TEA

Place broken sprigs or crushed leaves and water in large saucepan or pot. Bring to boil. Cover. Cut heat down. Simmer 15 minutes. Take off heat. Steep 10 minutes. Strain and sweeten to taste.

or

Pour boiling water over crushed dried or fresh leaves. Cover. Set aside to steep 15 minutes. Strain and sweeten to taste.

or

Put dried or fresh leaves in pot boiling water. Cover. Simmer 5 minutes. Steep 10 minutes or more. Strain and sweeten to taste.

NOTE: Use about 1 teaspoon crushed dried leaves or 2 tablespoons of crushed fresh leaves for each cup of water used.

A FEW EASY-TO-FIND PLANTS FOR BREWING NUTRITIOUS TEA IN THE WILDS:

ALFALFA: Dry and crush flower heads and young leaves.
BIRCH: Twigs. Nutritious but bland. Mix with others.
BLACKBERRY: Crumpled dry leaves. Loaded with vitamin C.
CHAMOMILE: Entire fresh or dried plant, flowers and/or leaves.
CLOVER: Dried flower heads. An extremely healthy tea.
COMFREY: Crushed Dried leaves.
MINT: Crushed fresh or dry leaves. Highly aromatic.
PEPPERMINT: Fresh or dried leaves. Highly aromatic.
PERSIMMON: Dried leaves. Loaded with vitamin C.
PINE: Fresh pine needles. Rich in vitamin A and C.

1. Light green needles in spring make best tea.
2. Older needles also bring good results.

PLANTAIN: Crushed fresh or dried leaves. One of best backwoods teas.
RED RASPBERRY: Dried leaves. Rich in vitamin C.
ROSE: Fresh or dried rose hips and fresh petals. Very rich in vitamin C.
SLIPPERY ELM: Small chips of slimy inner bark. Pleasant tasting and quite nourishing.

SPEARMINT: Fresh or dried leaves. Highly aromatic.
STINGING NETTLE: Young leaves and shoots. Good source of iron, protein and vitamin A and C.
STRAWBERRY: Dried leaves. Rich in vitamin C. Pleasant tasting.
SWEET VIOLET: Dried leaves. Loaded with vitamin A and C.
WINTERGREEN: Fresh or dried leaves. Delightfully aromatic.

PART III

EDIBLE PLANT IDENTIFICATION

(By words and pictures)

ALFALFA

Grows to 3" tall. Oval shaped leaves in groups of three. Tiny blue-violet clover-like blossoms (June to August). Widespread throughout U.S. (March to October).

AMARANTH, GREEN

Shallow bright red taproot. Stout, slightly hairy stem. Dull green egg-shaped leaves. Small greenish flower clusters. Commonly found in rich cultivated fields, yards, fence rows and waste ground. Troublesome weed. Widespread throughout U.S.

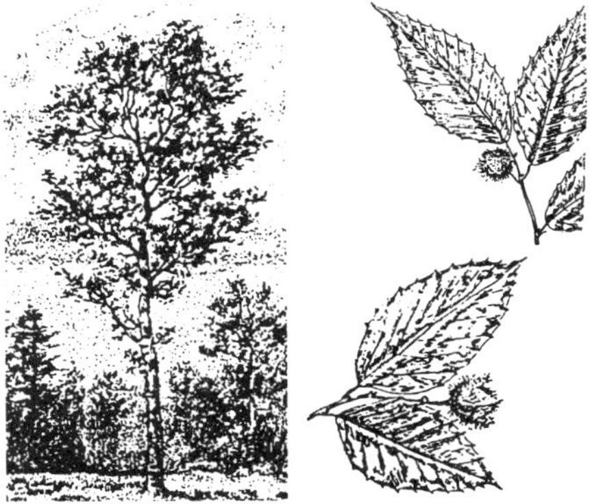

AMERICAN BEECH

Tree 45' to 60' tall. Bark dark gray with light gray patches. Leaves paper thin with coarse sharp teeth. Nuts in tiny burr with prickles. Ripe (September/October). Found in rich woods, uplands, moist rocky ground, etc. Common in many parts of U.S.

ASPARAGUS, WILD

Grows to 9' high. Lacy, dark green plant. Needle-like branches. Small green bell-shaped flowers followed by bright red berries (June) with black seeds. Found in gardens, fields, along ditches, meadows, orchards, etc. Widespread throughout U.S.

BIRCH

Tree to 60' tall. Simple saw tooth leaves. Sweet, aromatic black or gray papery bark peels in curls. Broken twigs and leaves have strong wintergreen smell. Found in rich woods, on river banks, moist fertile ground, etc. Common in many parts of U.S.

BLACKBERRY

Sprawling thorny shrub. Double toothed leaves. Showy white/pale pink 5-petaled flowers (April to July). Juicy purple/black berries (June to September). Found on wood edges, fields, road sides, along fences, sunny thickets, etc. Widespread throughout U.S.

BLUEBERRY (BILBERRY)

Grows to 2' tall. Narrow leaves with tiny teeth. Urn-shaped 5- petaled flowers (May to July). Berries (August to September) covered with white powder. Found in open woods, bogs, fields, thickets. Extremely adaptable. Widespread throughout U.S.

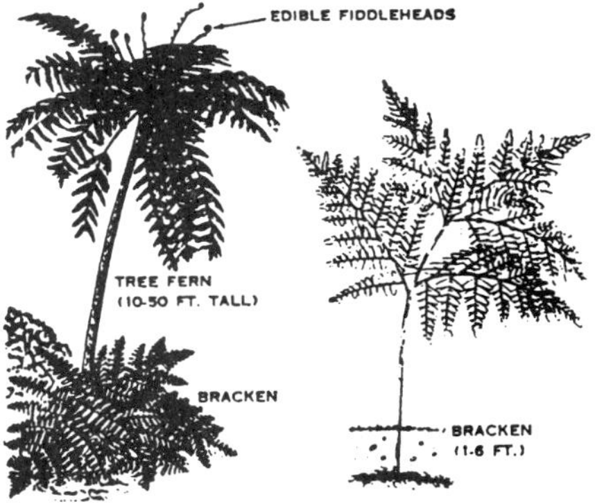

BRACKEN FERN

Grows 1' TO 6' high. Broad triangle-shaped leaves. Long creeping rootstalks. New ferns (fiddleheads) come out in spring. Found in open woods, thickets, waste places. Often found in areas with no other plants. Widespread throughout U.S.

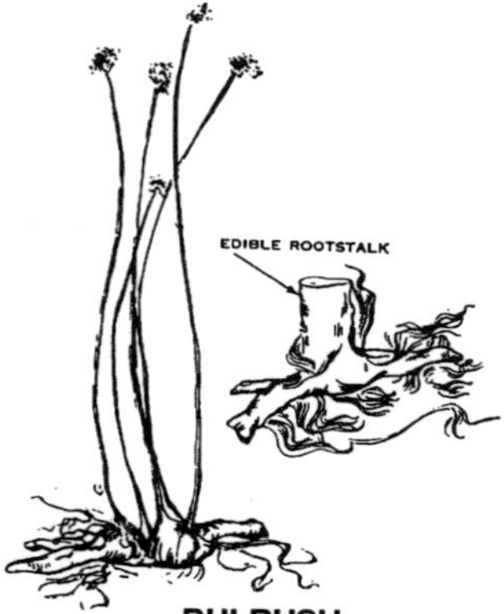

BULRUSH

Tall dark green water plant. Leafless, stiff round stems grow 3' to 9' high. Clusters of brown, bristly flower spikes loaded with seeds and pollen. Thick scaly root stocks. Found growing in thick patches in shallow water or mud. Widespread throughout U.S.

BUTTERNUT TREE

Grows 30' to 45' tall. Broad, spreading medium-size tree. Heavy lower branches. Scraggly limbs, sparse foliage. Very rough bark. Found in pastures, rich woods and on moist hillsides. Widespread throughout U.S.

BURDOCK

Grows 4' to 6' high. Resembles rhubarb. Hollow stems. Thick fleshy taproot. Egg-shaped deep green leaves. Clusters of purple flowers on short stalks (July to October). Found on neglected farm lands, fields, gardens, etc. Widespread throughout U.S.

CAMAS LILY (WILD HYACINTH)

Grows 1' to 2' high. Showy spikes of bright blue flowers (May through June). No branches or stems. Grass-like leaves. Found along streams, in wet fields, meadows, moist woods. Widespread throughout U.S.

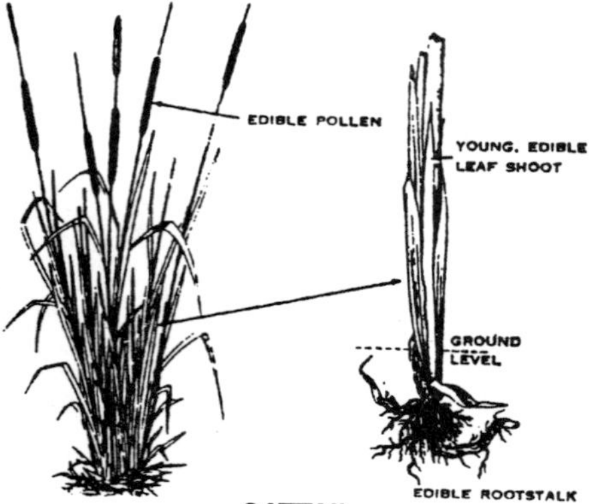

CATTAIL

Grows to 8' high. Spongy, long sword-like leaves. Yellow pollen laden flowers on erect stalks (May to July). Straight stems with no branches have hotdog-like seed heads. Found in marshes, ponds, shallow water, ditches, etc. Widespread throughout U.S.

CHAMOMILE

Grows to 2' high. Bright green stemless leaves. Delicate apple or pineapple smell. Daisy-like, yellow-centered, tight budded flowers with white petals (May through October). Found almost everywhere -- meadows, alongside roads, etc. Widespread throughout U.S.

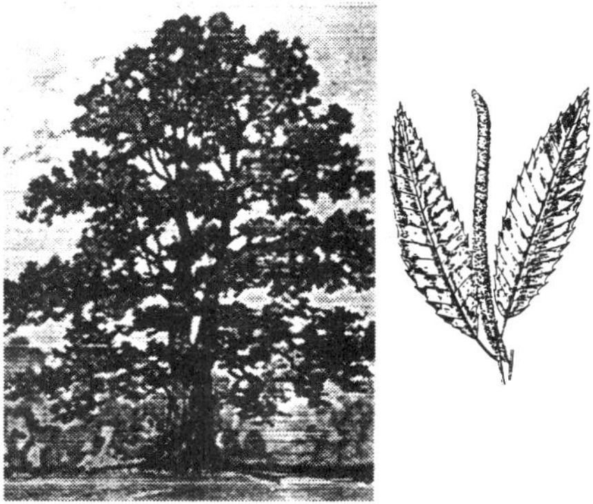

CHESTNUT

Tree grows from 50'to 70' tall. Large spear-shaped, coarsely toothed leaves. Fruit,is 2" to 3" bur containing 3 nuts. Yellow-green flowers (July). Found in well drained forests, rocky woods, pastures, etc. Common in many parts of U.S.

CHICKWEED

Succulent, straggling stems up to 1' long. Hairy leaf stalks creep on ground. Oval, pointed leaves. Small white star-shaped flowers (March to September). Found in fields, gardens, sides of road, waste places and cultivated grounds. Widespread throughout U.S.

CHICORY, WILD

Grows 1' to 4' high. Deep carrot-like root. Red veined leaves at plant base like the dandelion. Milky sap. Hollow stems rough and hairy. Abundant flowers. Found in fields, on sides of roads, fence rows, lawns, pastures, etc. Widespread throughout U.S.

CLOVERS

Grows to 18" high. Pink to red to purple rounded heads (May to September). Found everywhere -- fields, lawns, parks, roadsides, etc. Widespread throughout U.S.

COMFREY

Grows to 3' high. Rough and hairy. Stout, hollow stem. Large hairy leaves cause itching when skin is touched. Bell-shaped flowers (May to June). Found in wet soils -- river banks, ditches, roadsides, meadows, etc. Widespread throughout U.S.

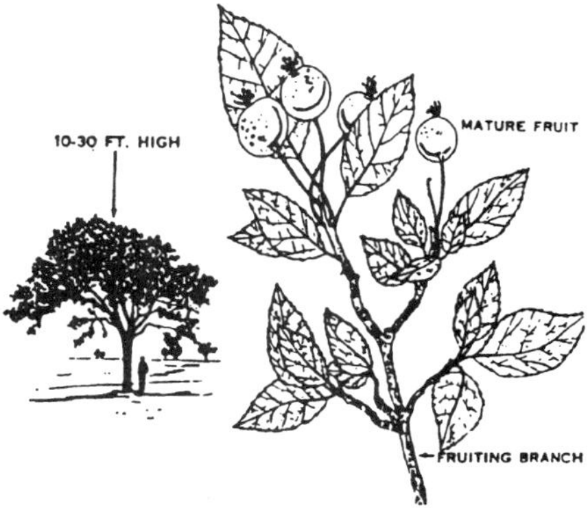

CRAB APPLE, WILD

Small tree. Leaves and fruit on short twigs with thorny ends. Short pink-white flowers with 5 petals (April to June). Tiny apples 1" across (September to November). Found in open woodlands and edge of woods, yards, etc. Widespread throughout U.S.

DAISY OX-EYE

Grows to 2' high. Numerous flowers with oblong, white petals around yellow pebbly center. Leaves narrow, oblong and deeply scalloped. Found in meadows, waste areas, fields, on side of roads, etc. Widespread throughout U.S.

DANDELION

Grows to 2' high or more. Milky juice in hollow stalk. Thick taproot. Jagged dark green, hairless leaves clustered near ground. Golden yellow flowers on leafless stem. Found everywhere -- lawns, parks, gardens, roadside, fields, etc. Widespread throughout U.S.

DAY LILY

Grows from 2' to 5' high. 3" to 5" showy orange flowers on top of naked stem open only one day. Frequently grows in dense patches. Found along roadsides and ditches in large clumps. Widespread throughout U.S.

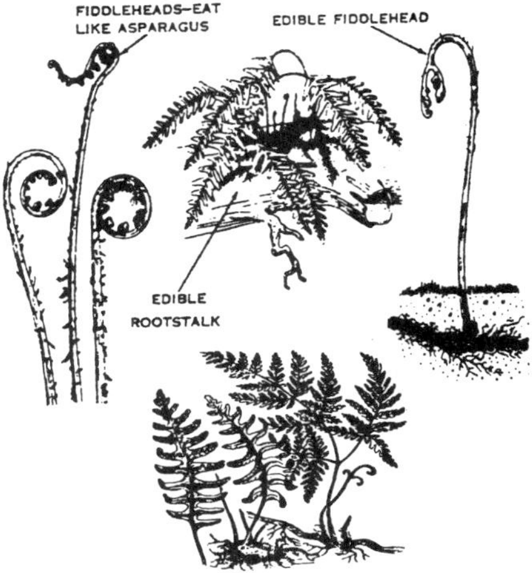

EDIBLE FERN PARTS

Widespread throughout U.S.

ELDERBERRY

Grows to 12' high. Thick, white pith in stems and twigs. Fragrant white flowers in bowl-shaped clusters (June to July). Tiny purplish berries (July to September). Found in ditches, thickets, along roads, river banks, etc. Widespread throughout U.S.

EVENING PRIMROSE

Grows to 5' high. Long carrot-like root. Reddish stem. Tall leafy spike with large yellow flowers. Subtle fragrance. Flowers open in late afternoon. Found on roadsides, in fields, sand dunes and waste places. Widespread throughout U.S.

FIREWEED (PILEWORT)

Grows 3' TO 8' high. Erect unbranched stems. Narrow willow-like leaves. Showy rose-purple flower spikes. Grows thickly in burned over areas, logging sites and on river banks. Widespread throughout U.S.

GALINSOGA (QUICKWEED)

Very small white flowers surround small yellow stems with many branches. Leaves pointed at tip. Found in weedy gardens, lowland fields, waste places -- especially damp areas with rich soil. Widespread throughout U.S.

GOOSEGRASS (CLEAVERS)

Straggly stems grow to 4' high. Tiny star shaped greenish-white flowers. Stems and leaves covered with rough bristles. Found in moist woods, thickets, roadsides, waste places, fields, etc. Widespread throughout U.S.

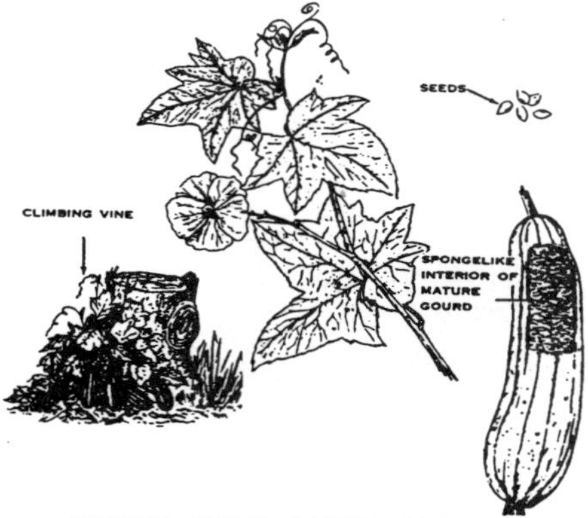

GOURD, WILD (LUFFA SPONGE)

Found on a climbing vine 20' to 30' long. Yellow flowers. Member of squash family. Grows similar to watermelon, cantaloupe and cucumber. Often found in old gardens, clearings and barnyards. Widespread throughout U.S.

GRAPE, WILD

High climbing vine. Shreddy, brown bark. Heart-shaped leaves. Greenish flowers (August to October). Fleshy grapes in clusters. Found in gardens, thickets, edges of woods, etc. Check grape smell since there is poisonous look alike. Widespread throughout U.S.

GREENBRIER

Vine with scattered thorns. Triangle-shaped leathery leaves. Greenish flower clusters. Round black berries follow flowers. Found on sides of roads, old fields and open woods and thickets. Widespread throughout Midwest and Eastern U.S.

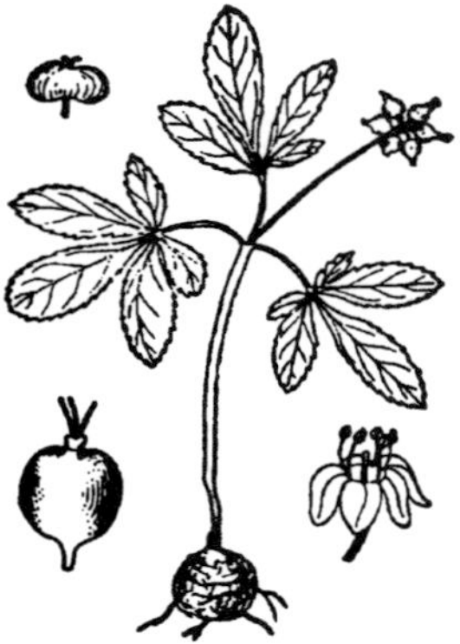

GROUNDNUT

Slender vine often found climbing on other plants. Climbs to height of several feet. Purplish-brown highly fragrant flowers. Milky sap in plant. Found in rich thickets and wet areas, around ponds and along streams. Widespread throughout U.S.

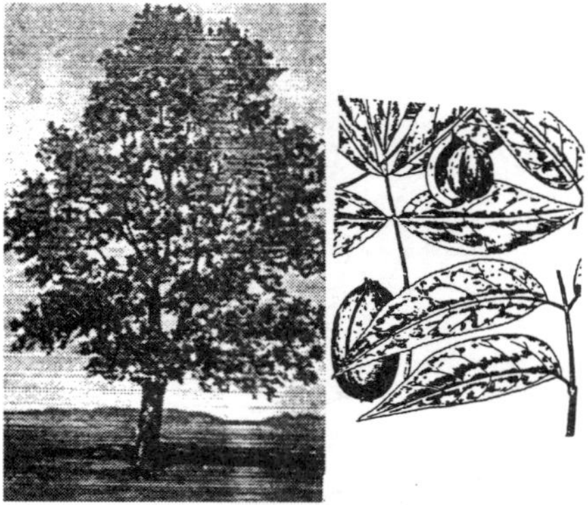

HICKORY

Grows 60'to 90' tall. Deep yellowish green leaves from 8" top 14". Trunk 4' diameter with light colored bark. Egg-shaped nuts with thick husks. Extremely hard wood. Found in rich soil, river bottoms, hillsides, etc. Common in many parts of U.S.

HOG PEANUT

Low growing vine with slender stems. Light green leaves. Clusters of white to pale lilac flowers. Likes rich soil, thickets, wet or moist wooded areas. Widespread throughout U.S.

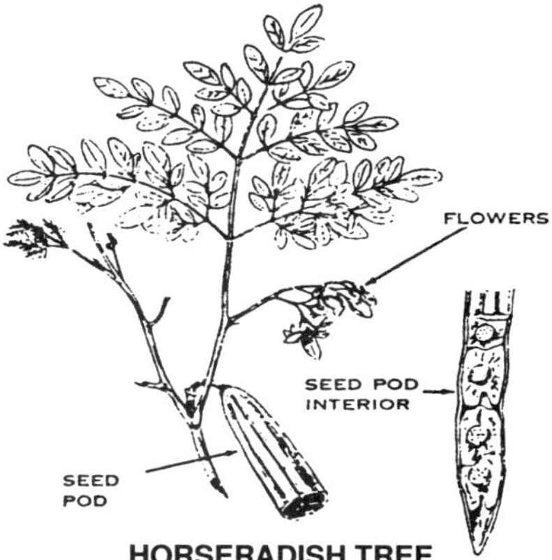

HORSERADISH TREE

Grows 15' to 45' tall. Flowers on branch ends as are long pendulum fruit resembling giant beans. Leaves fern-like. Found in abandoned fields, gardens, edges of forests, etc. Widespread throughout U.S.

HUCKLEBERRY

Narrow leaves with tiny teeth. Hairy twigs and branches. Bell-like white or pink or green flowers (May to July). Blue-black berries follow flowers. Found in swamps, bogs, moist woods, etc. Often grows near blueberries. Widespread throughout U.S.

HYSSOP HEDGE-NETTLE

Small erect shrub. Square stems grow to 2' long. Dense flowers on top of stem. Found in wet places, usually sandy soil, meadows, fields, shores, etc. Widespread throughout U.S.

INDIAN CUCUMBER ROOT

Grows to 2' high. Yellow-green flower cluster at stem top. Inedible purple berries follow flowers. Underground swollen white tuber. Found on swamp edges and in moist woods. Widespread throughout Eastern U.S. and as far west as Kansas.

LADY'S THUMB (REDLEG)

Leaves vary -- some narrow, others broad. Flowering spikes. Flowers purplish, green and pink. Found in cultivated fields, roadsides, in ditches, damp clearings. Widespread throughout U.S. except for some of southwest.

LAMB'S QUARTER (PIGWEED)

Grows 1' to 5' high. Short taproot. Triangular dark green leaves covered with mealy white powder. Irregular spike clusters at branch ends. Flowers small and green. Commonly found growing in gardens and grain fields. Widespread throughout U.S.

LEEK, WILD (RAMP)

Grows to 18" high. Slim fleshy leaves (2 or 3). Whitish bulbs exude strong onion smell. Spoke-like cluster of creamy white flowers (May to June). Found in rich moist soil, wet woods, alongside roads, thickets, etc. Widespread throughout U.S.

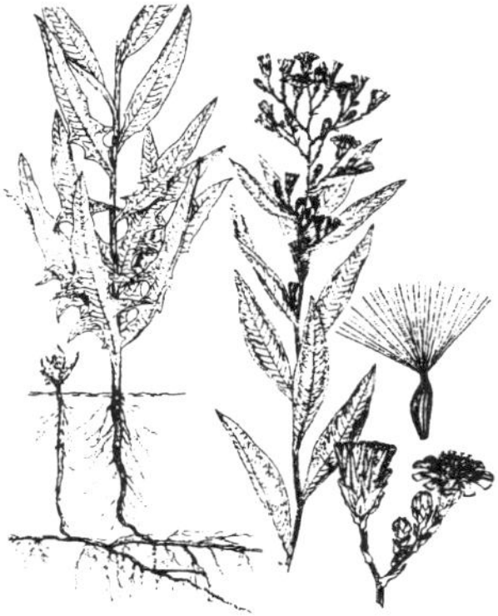

LETTUCE, WILD

Grows 1' to 5' high. Rounded blue-green leaves similar to dandelion. Strong odor. Prickly stem. Small blue or yellow flowers. Broken parts exude milk-like juice. Found on roadsides, banks, vacant lots, waste areas, etc. Widespread throughout U.S.

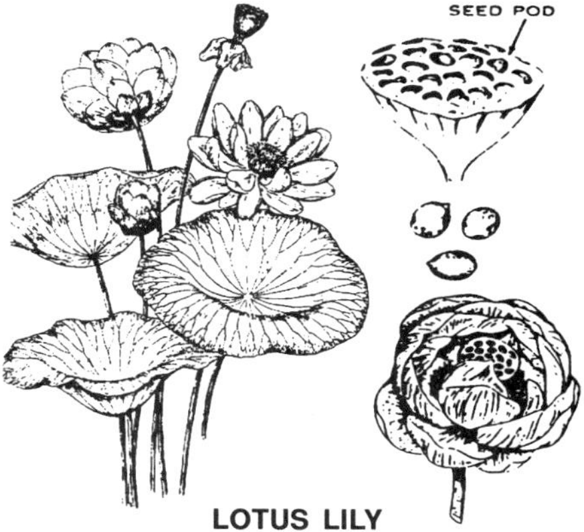

LOTUS LILY

Grows 5' to 6' high. Aquatic plant. Leaves measure 3' to 5' wide. Thick tubers grow 50' long. Yellow or pink flowers are 4" or more across. Found in quiet pond or lake water, along sluggish rivers, etc. Common in many parts of U.S.

MALLOW, COMMON

Grows to 14" high or more. Deep rooted plant. Low growing trailing stems. Rounded, slightly toothed leaves. Small white or lilac flowers (April to October). Found in yards, gardens, cultivated fields, barnyards, etc. Widespread throughout U.S.

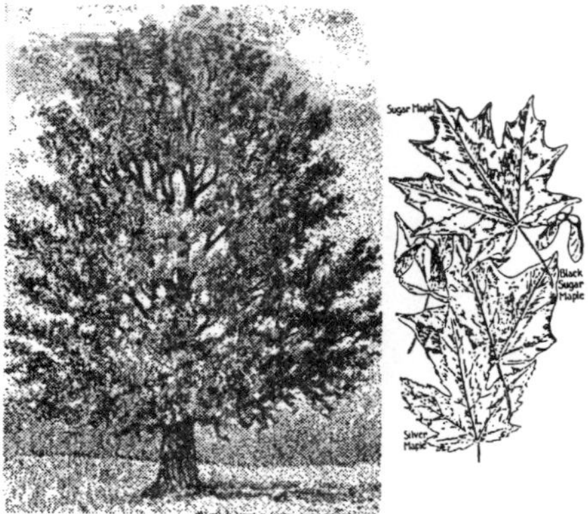

MAPLE TREE

Grows from 50' to 70' high with 5' trunk. Serrated 3-sectional leaves. Light brown/ gray bark. Some bark smooth, some rough. Gray lichen moss often found on these trees. Common in rich woods, rocky hillsides, yards, etc. Widespread throughout U.S.

MARSH-MARIGOLD (COWSLIP)

Aquatic plant. Succulent hollow stem. Kidney-shaped leaves. Large buttercup-like flowers (May to July). Grows around swamps, ponds, lakes, in wet ditches, etc. Common in many parts of U.S.

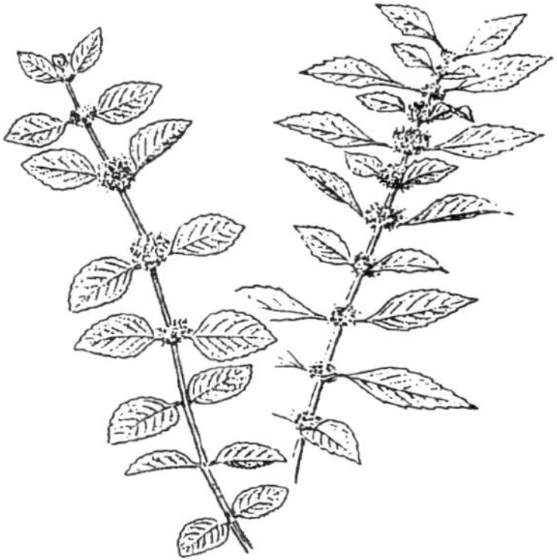

MINT

Grows to 3' high. Fine hairs on stem. Strong mint aroma. Tiny clusters bell-shaped flowers (June to October). Found in wet or damp places such as roadsides, beside streams, around ponds, etc. Widespread throughout U.S.

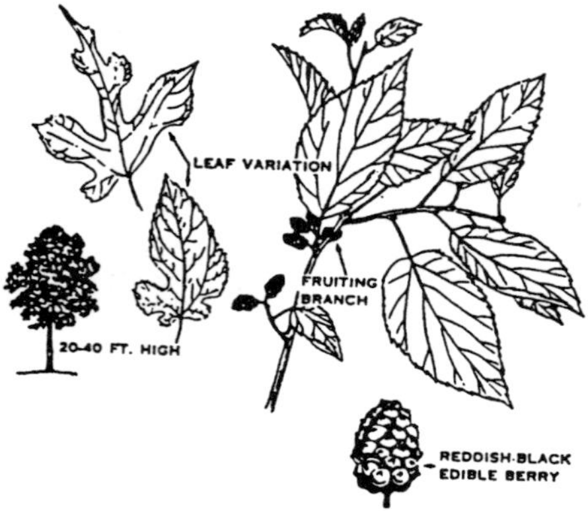

MULBERRY, RED

Grows 20' to 60' tall. Heart-shaped sandpapery leaves. Milky sap in twigs. Flowers in tight drooping clusters (April to May). Fruit resembles blackberries. Found along roads, in abandoned fields, open woods, thickets, etc. Widespread throughout U.S.

MUSTARD, WILD

Grows to 10' high. Rounded, coarse and bristly leaves. Bright yellow flowers in clusters at branch ends (May to September). Pods contain black seeds. Found in hedges, waste places, cliffs, roadsides, etc. Widespread throughout U.S.

NETTLE, STINGING

Grows to 5' high. Coarse saw-toothed leaves. Erect square stems. Covered with stiff stinging hairs. Green flower clusters on many spikes (June to September). Found in vacant lots, gardens, ditches, along sides of roads, etc. Widespread throughout U.S.

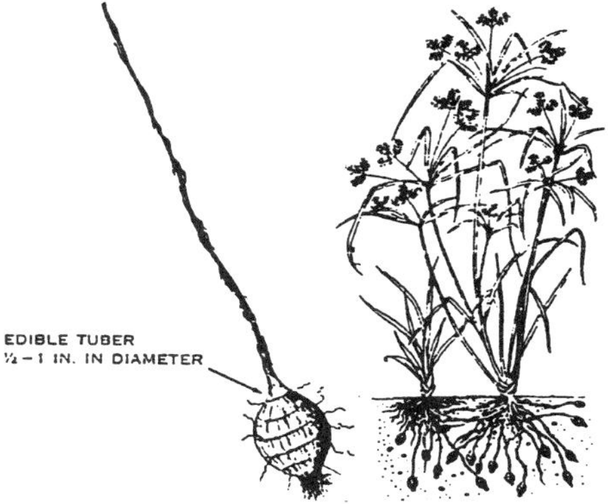

NUT GRASS (CHUFA)

Clump of grassy-looking leaves on triangular stem. Flowers in clusters of feathery spikelets (June to October). Found in moist, sandy places, edge of streams and ponds, and moist or wet ground. Widespread throughout U.S.

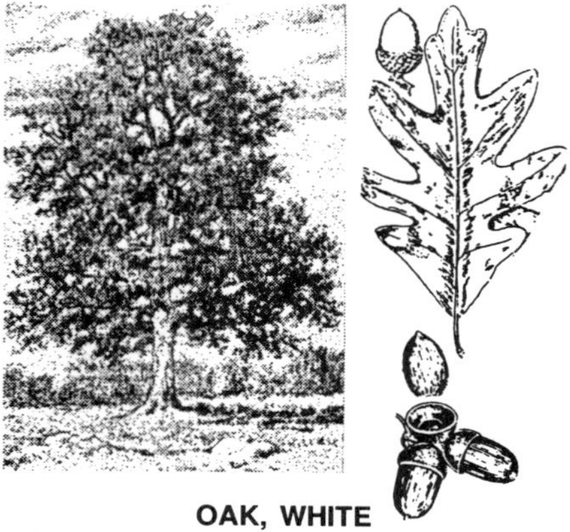

OAK, WHITE

Grows to 110' tall. Bright olive-green leaves to 7" long. Light brown, sweet, edible acorns grow in pairs. Flowers (May and June). Found in dry woods, sandy places, gravelly ridges, etc. Common in many parts of U.S.

ONION, WILD

Grows to 2' high. Grass-like leaves in early spring. Small pink or purple flowers form clusters on stem end (June to August). Found on slopes, ridges, open woods, hill-sides, etc. One of the most abundant food plants. Widespread throughout U.S.

OSTRICH FERN

Grows from 2' to 6' high in graceful bunches. Rich looking, dark green fronds (fern leaves). Found on banks of rivers and streams, around swamps, clear areas of woods, in rich, wet dirt. Common in many parts of U.S.

PARSNIP, WILD

Grows to 4' high. Erect flowering stalk. Fruits have flat narrow wings. Commonly found on roadsides, in ditches, grassy waste places, etc. Widespread throughout U.S.

PENNYCRESS, FIELD

Grows 16" to 32" high. White flowers (April to August). Deeply notched seedpods. Found in waste places, gardens, grasslands, fields, roadsides, etc. A troublesome weed in grain fields. Widespread throughout U.S.

PEPPERGRASS

Grows from 1' to 2' high. Leaves on stem are arrow-shaped. Flower clusters on top of plant. Found on roadsides, fields of clover, alfalfa and winter wheat. Widespread throughout U.S.

PEPPERMINT

Grows to 3' high. Dark green, sharply toothed leaves. Mint smell. Taste initially burns, followed by cool sensation. Small lavender blossoms in spike-like groups (July to September). Found on brook and river banks, wet meadows, etc. Widespread throughout U.S.

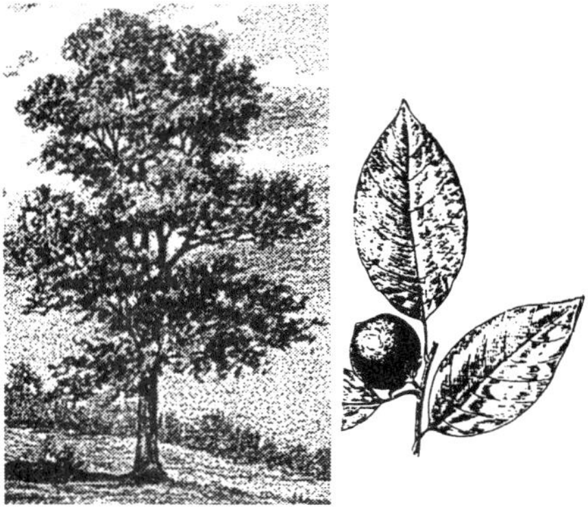

PERSIMMON TREE

Grows to 60' tall. Stiff, glossy, dark green leaves. Fruit orange to reddish-purple when ready to eat (August to October or later). Found in old fields, open woods and in dry non-evergreen woods. Common in many parts of U.S.

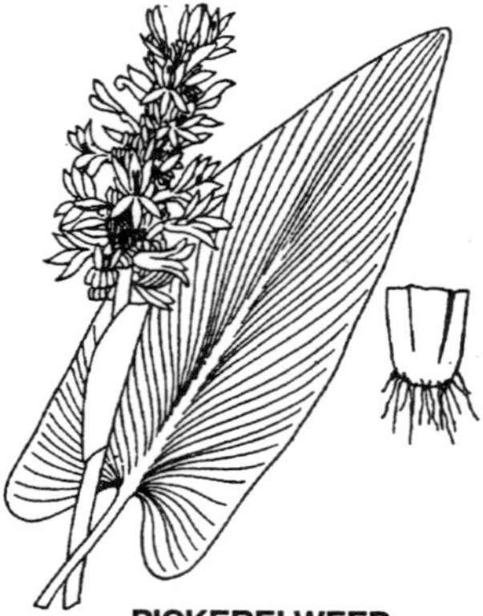

PICKERELWEED

Aquatic plant. Grows to 3' high. Arrow-head-shaped leaves. Dense blue flowers. Found in marshes, on banks of slow moving streams, around lakes and ponds. Common in many parts of U.S.

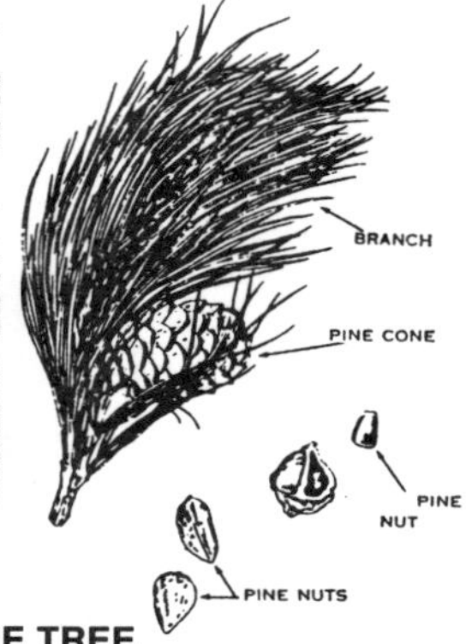

PINE TREE

Grows to 150' tall. Soft blue-green needles, 4" long or longer. Needles in clusters of 2 to 5. Crushed needles have characteristic taste and smell. Dark brown cylindrical pine cones, 4" to 8" long. Common in many parts of U.S.

PLANTAIN, BROADLEAF

Grows to 18" high. Broad, oval, toothed leaves. Compact clusters of tiny yellow or greenish flowers on leafless stems (May to August). Found in lawns, fields, pastures, waste areas and gardens. Widespread throughout U.S.

POKEWEED (POKE)

Grows 2' to 3' high. Smooth reddish stems. Plant has disagreeable smell. Small flower clusters greenish to white. Found in pastures, cultivated fields, roadsides, open places in woods, etc. Common in many parts of U.S.

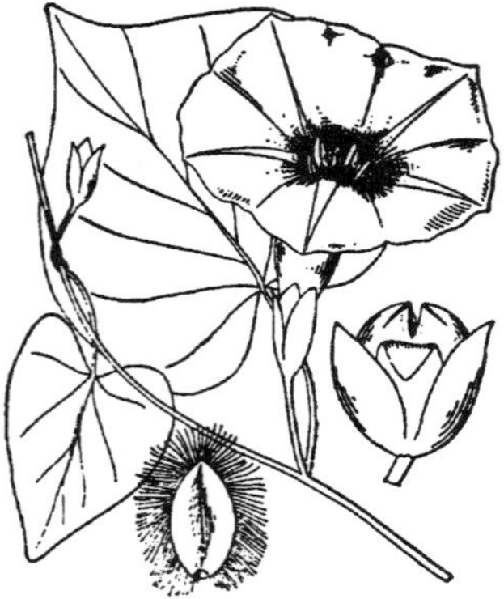

POTATO VINE, WILD

Trailing vine. Grows to 12' long. Heart-shaped leaves. Large white funnel-shaped flowers with pink-purple centers. Large root resembles sweet potato. Found in fields, pastures, on roadsides, etc. Widespread throughout U.S.

PRICKLY PEAR CACTUS

Grows 1' to 8' high. Tufts of bristles. Sharp spines. Showy yellow flowers (May to August). Fruit dull red, pulpy (August to October). Found in dry, sandy soils, rocks, etc. Common in many parts of U.S.

PURSLANE

Grows to 12" high. Creeps on ground. Reddish stems. Thick, succulent, spatula-shaped bright green leaf clusters. Tiny yellow flowers open only on sunny mornings (June until frost). So common that all you have to do is look for it. Widespread throughout U.S.

QUEEN ANNE'S LACE (WILD CARROT)

Grows to 4' high. Hairy oblong leaves and stems. Lacy clusters of purple-pink flowers (June to September). Leaves and flowers smell like carrots or plant may be poisonous look alike. Found in meadows, fields, pastures, roadsides, etc. Widespread through U.S.

RED BUD TREE

Grows to 40' tall. Heart-shaped leaves. Showy clusters of purple-red, pea-like flowers (March to May). Found on stream borders, roadsides, in woods, yards, etc. Widespread throughout U.S.

RED RASPBERRY (CLOUDBERRY, DEWBERRY, ETC.)

Grows to 5' high. Prickly shrub. White powder on stems. Tiny white flower clusters (April to August). Red berries. Found along roadsides, edges of woods, fields, etc. Widespread throughout U.S.

RHUBARB

Grows 1' to 4' high. Large, compact plant. Thick stems. Leaves 2' to 4' long. Peculiar aromatic odor. Greenish-white flower clusters (May to August). Found everywhere -- parks, roadsides, fields, yards, etc. Widespread throughout U.S.

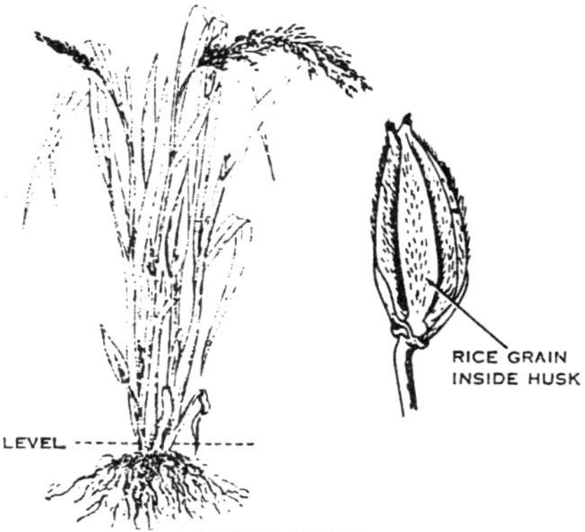

RICE, WILD

Grows 3' to 4' high. Aquatic grass. Coarse and graceful. Light green 2" wide leaves. Large plumes of flowers. 2' flower stems. Found in wet areas, edges of ponds, rivers and lakes -- in shallow water. Common in many parts of U.S.

ROSE, MULTI-FACETED WILD

Grows to 10' high. Look for fruit -- fleshy red hips. All rose varieties have thorns or briars. Found on edges of woods, old pastures, yards, sides of roads, etc. Widespread throughout U.S.

SERVICEBERRY

Small tree or shrub. Oval leaves with blunt tips and tiny teeth. White flowers in drooping clusters (May to July). Fruit dark blue and much like huckleberries (June to September). Found in thickets, open woods, fields, swampy areas, etc. Widespread.

SHEEP SORREL (COMMON SORREL)

Grows to 18" high. Stems erect, slender and branched at top. Hard to eradicate. Arrow-shaped leaves have pleasant, mildly sour taste. Flowers yellow to red. Found in fields, gardens, pastures, roadsides, meadows, lawns, etc. Widespread throughout U.S.

SHEPHERD'S PURSE

Grows to 4' high. Dandelion-like, hairy leaves in clusters. Hairy stems. Triangular seed pods. Small erect spikes of tiny white flowers (January to December.) Found on roadsides, in backyards, gardens, fields, etc. Widespread throughout U.S.

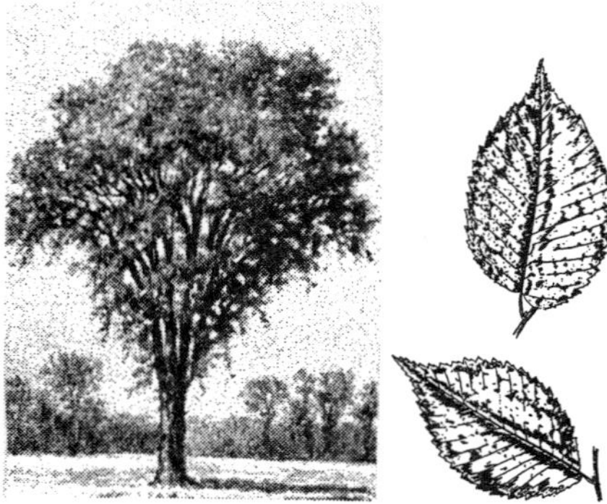

SLIPPERY ELM TREE

Grows to 50' feet. Rough bark. Mildly aromatic. Oval leaves with sandpapery tops. Fruits winged. Ripen (March to May). Flowers bloom (March to April). Found in moist woods, thickets, rocky hillsides, etc. Common in many parts of U.S.

SOLOMON'S SEAL

Grows to 3' high. 4" x 2" dark green leaves. Drooping clusters of fragrant, greenish-white flowers (April to May). Black or blue berries. Found in rocky woods, shady places, dry to moist thickets, fields, etc. Widespread throughout U.S.

SPRING BEAUTY

Grows 6" to 12" high. Pair smooth leaves halfway up stem. Flowers white to pale pink with dark pink veins (March to May). Solid bulb-like root buried 3" to 5" inches. Found in moist woods, rich soil, thickets, etc. Widespread throughout U.S.

STRAWBERRY, WILD

Similar to cultivated strawberries. Smaller berries. Flat clusters of white flowers. Leaves dark green and hairy. Found in moist, rocky areas, clearings in woods, fields and other open places. Some in swampy areas. Widespread throughout U.S.

SUNFLOWER

Grows 3' to 12' high. Stems are stout, coarse, rough and hairy. Saw-tooth leaves. Showy bright yellow flowers are 3" to 5" across. Found on open plains, cultivated fields, waste places, pastures, grain fields, roadsides, etc. Widespread throughout U.S.

SWEET FLAG (CALAMUS)

Grows to 4' high. Cattail-like leaves have spicy smell when bruised. Stalks covered with tiny green-yellow flowers (May to August). Found in shallow water, swamps, wet fields, stream edges, etc. Widespread throughout U.S.

SYCAMORE TREE

Grows 50' to 80' tall. Massive tree with 12' diameter trunk. Dull brown brittle and flaking bark. Coarsely toothed leaves 6' to 10' long. Small flowers (April to May). Tiny hairy fruit. Found on banks of streams, bottom lands. Widespread throughout U.S.

THISTLE, BULL

Grows to 8' high. Dozens of varieties. Long needle pointed barbs on leaves and stems. Leaves often white and wooly on underside. Many white, red or purple burr-like flowers. An aggressive weed. Found in pastures, roadsides, etc. Widespread.

VIOLET, SWEET

Grows 4" to 6" high. Green heart-shaped leaves with scalloped edges. Sweet smelling, pansy-like flowers (May to June). Found on playgrounds, damp shady woods, yards, moist soil on stream edge, etc. Widespread throughout U.S.

WAPATO (DUCK POTATO)

Grows 10' or more. Aquatic plant. Leaves shaped like arrowheads. Waxy white flowers near tops of stalks (July and August). Smooth potato-like tubers on ends of underground runners. Found on stream edges and road borders, swampy areas. Widespread.

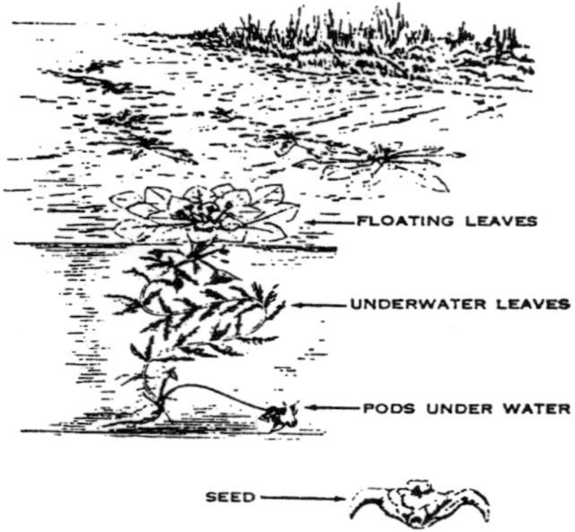

WATER CHESTNUT

Aquatic free floating plant. Some leaves on top of water. Others, root-like and feathery are underwater. Pods underwater. Nuts resemble horned steer. Found on lakes, ponds, rivers in quiet water. Common in many parts of U.S.

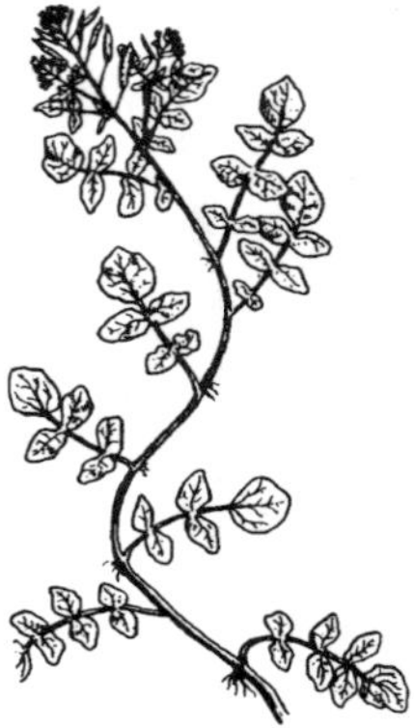

WATERCRESS

Aquatic plant Succulent creeping stems. Grows underwater or partly submerged in water. Shiny dark green leaves. Tiny white flower petals. Found in slow moving streams, springs, ponds, lakes, etc. Widespread throughout U.S.

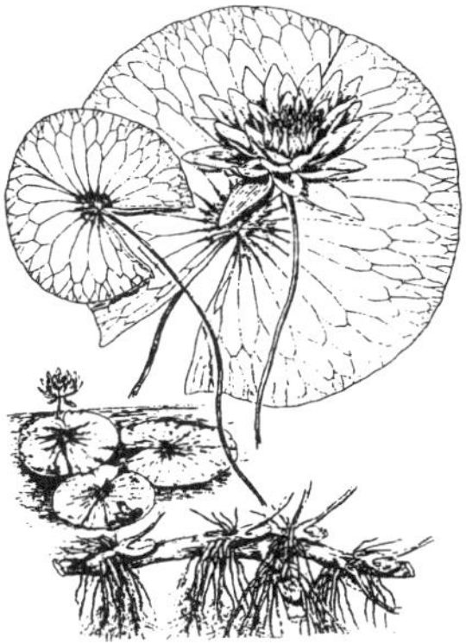

WATER LILY, FRAGRANT

Aquatic plant with large 8" to 28" floating leaves. Showy sweet smelling pale yellow flowers 5" to 10" across (June to September). Found in quiet or dead water -- ponds, slow running brooks, lakes, etc. Widespread throughout U.S.

WATER PLANTAIN

Heart shaped leaves. Thick bulb-like root stocks. White or yellow flowers. Plant stays partially submerged in water. Found in lakes, ponds and slow moving streams. Common in many parts of U.S.

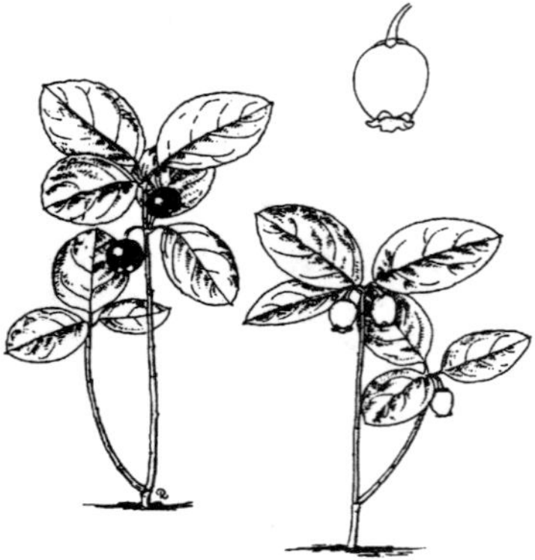

WINTERGREEN

Grows to 6" high. Low creeping evergreen. Aromatic. Glossy, oval leaves. Waxy white flowers like drooping bells (June to August). Dry red berries (fall and winter). Found in dry, wooded areas, clearings, base of trees, etc. Widespread throughout U.S.

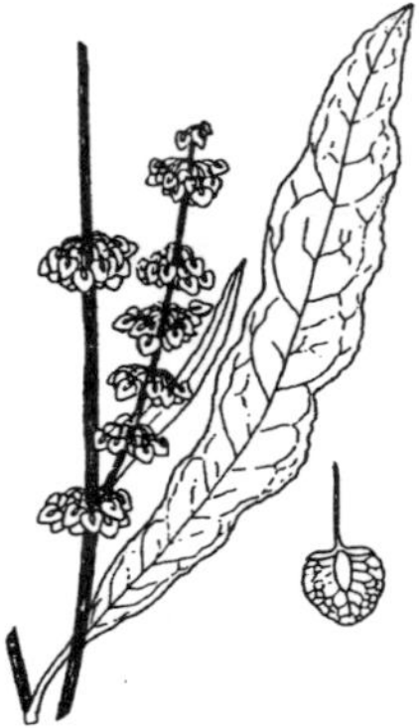

YELLOW DOCK (CURLY DOCK)

Grows to 5' high. Deep yellow root. Lance shaped leaves. Winged, heart-shaped seeds (June to September). Small greenish flowers densely clustered on stalks (May to September). Found in pastures, meadows, gardens, roadsides, etc. Widespread in U.S.

PART IV

WHAT NOT TO EAT -- POISONOUS PLANTS

9

Poisonous Plants and How to Avoid Eating Them

1. Poisonous plants are those containing natural toxic substances.
2. Poisonous plants aren't a serious hazard. Their number is small compared to the number of edible nonpoisonous plants.
3. Under certain conditions a person can be poisoned by:
    a. Eating poisonous plants or plant parts.
    b. Absorbing plant poison through the skin.
    c. Inhaling the plant poison.
4. Mistakenly eating poisonous plants can be extremely dangerous. They can cause:
    a. Serious illnesses.
    b. Death.
5. The critical factor in using plants for food:
    a. Eat only plants known for certain to be safe.

Learn All You Can About Poisonous and Nonpoisonous Plants

1. Preparation for survival situations includes learning to readily identify poisonous plants.
   a. Recognizing poisonous plants is as important as recognizing edible plants.
   b. Successful use of plants for food in a survival situation depends entirely upon positive identification.
2. Recognizing poisonous plants will help avoid unnecessary problems resulting from eating them:
   a. Remember: poison hemlock has killed people who mistook it for wild carrots and wild parsnips.
3. There is no room for experimentation where plants are concerned. There may be no second chance.
4. Many deadly poisonous plants:
   a. Look like their safe to eat relatives
   b. Have other edible look-a-likes in the plant world.
   c. For example, poison hemlock is similar to Queen Anne's Lace (wild carrot).

5. A few plants are safe to eat in certain stages of growth and poisonous in other stages:.
   a. For example, pokeweed leaves are edible when it first starting to grow, but they soon becomes poisonous.
6. A person can eat the fruit of some plants only when ripe.
   a. For example, the ripe fruit of mayapple is edible, but the green fruit is poisonous.
7. Some plants contain both edible and poisonous parts.
   a. Tomatoes and potatoes are commonly known plant foods, but their green parts are poisonous.
8. Some plants become poisonous after they wilt.
   a. For example, when the black cherry wilts, hydrocyanic acid is developed.
9. Specific preparation methods make some plants edible that are poisonous when eaten raw.
   a. A person can eat dried corms of the Jack-in-the pulpit.
   b. The same corms are poisonous if not thoroughly dried.

How Poisonous are Poisonous Plants?

1. Poisons in plants exist in great number and variety.
2. Phytotoxins are the deadliest of poisons. These include:
   a. Abrin found in the Rosary Pea.
   b. Jatrophin found in the Physic Nut.
   c. Ricin found in the Castor Bean.
3. Ricin, for example, is:
   a. 12,000 times more potent than rattlesnake venom.
4. How poisonous is a poisonous plant?: It's impossible to tell how potent a poisonous a plant may be. It's equally as hard to tell how the particular plant will affect an individual who has eaten it. A number of factors must be considered:
   a. Every person has a different level of resistance to poisons. b. Some people are more sensitive than others to the poisons in a particular plant.
   c. The severity of poisoning in a person is influenced by the quantity of the poison ingested and by health and age.
   d. Poisonous plants vary in their amount of toxity due to climate, habitat, soil, season, age, etc.

e. A person must eat a lot of some poisonous plants before experiencing anything adverse.
f. Other poisonous plants will cause a person's death after eating only small amounts.
g. Poison may be distributed throughout a plant or may be concentrated in seeds, fruit, leaves or the roots.

Common Misconceptions About Poisonous Plants

1. Watch what animals eat and eat what they eat.
   a. This statement is not always accurate. Some animals can eat plants that would be poisonous to humans.
2. Boil the plant and poisons will disappear.
   a. Boiling does remove some poisons, but not always.
3. All red colored plants are poisonous.
   a. Some red plants are poisonous, but certainly not all of them.
4. The most important point is this:
   a. Make an effort to learn as much about plants as possible.
   b. Never eat a plant unless it's positively identified.

Signs and Symptoms of Plant Poisoning

1. Signs and symptoms of plant poisoning may include:
   a. Severe abdominal cramps.
   b. Depressed heartbeat.
   c. Slower than normal respiration.
   d. Dry mouth.
   e. Nausea and/or vomiting.
   f. Diarrhea.
   g. Headaches.
   h. Hallucinations.
   i. Unconsciousness.
   j. Coma.
   k. Death.

What to do When Someone is Poisoned by Eating a Plant

1. Act quickly and pull poisonous material out of victim's mouth.
2. Try to induce vomiting:
   a. Tickle back of person's throat.
   b. Make person drink warm salt water, if conscious.

3. Dilute poison by making person, if conscious:
   a. Drink large quantities of milk.b. Drink large quantities of water.

Plants and the Universal Edibility Test

1. When unsure as to how safe a plant may be, use the Universal Edibility Test. Apply test before eating. This is a reliable way to determine which plants to eat and which plants to avoid.
2. Before testing a plant for edibility, be sure there are enough plants around to make testing worth the time and effort.
   a. Each part of a plant (roots, flowers, leaves, etc) requires more than 24 hours to test.
3. To avoid potential danger, stay away from unidentified plants having:
   a. A bitter or soapy taste.
   b. Discolored or milky sap.
   c. Carrot, dill, parsnip or parsley-like foliage.
   d. Almond smell in woody parts and leaves.
   e. Fine hairs, thorns and spines.
   f. Pods with bulbs, beans or seeds.
   g. Pink, purplish or black spurs on grain heads.

4. Use the above criteria as eliminators when choosing plants for the Universal Edibility Test. These guidelines will help a person avoid eating dangerous toxic plants.

The Universal Edibility Test for Plants Found in the Wilds.

1. Don't eat for eight hours prior to taking the test.
2. Separate the plant into its basic components:
   a. Leaves.
   b. Buds and/or flowers.
   c. Stems.
   d. Roots.
3. Test only one part of a plant at a time.
4. Smell the plant for strong or acid odors.
   a. Keep in mind that smell alone doesn't indicate a plant is safe to eat.
5. Select small portion of a single plant:
   a. Prepare it the way you plan to eat it.
6. During test period:
   a. Drink nothing other than purified water.
   b. Eat nothing more than plant being tested.

7. Before placing the cooked plant part in mouth:
   a. Touch a small portion (a pinch) to outer lip.
   b. Check for any burning or itching sensation.
8. If after 3 minutes there's no reaction on lip:
   a. Place plant part on tongue.
   b. Leave there 15 minutes.
9. If still no reaction:
   a. Thoroughly chew tiny piece.
   b. Hold in mouth for 15 minutes.
   c. Do not, under any circumstances, swallow.
10. If no itching, stinging, burning, numbness, dizziness, etc., occurs:
    a. Swallow the plant piece.
11. Now wait 8 more hours. If ill effects occur:
    a. Induce vomiting.
    b. Drink plenty of water.
12. If no ill effects occur:
    a. Eat 1/4 cup of same plant part, prepared same way.
    b. Wait another 8 hours.
13. If no ill effects occur:
    a. The plant part as prepared is safe for eating.

14. Test every part of plant for edibility.
    a. Some plants have both edible and inedible parts.
    b. Never assume a part found edible when cooked is also edible when raw.
    c. Test the raw part before eating to ensure edibility.
    d. The same part or plant may produce varying reactions in different individuals.

PART V

IDENTIFICATION OF COMMON POISONOUS PLANTS

(By words and pictures)

AZALEA

TOXIC PART: Entire Plant.
SYMPTOMS: Slow pulse; vomiting; low blood pressure; watering of the nose, mouth and eyes; paralysis, convulsions, death.
DESCRIPTION: Shrub. Grows to 2'. Showy masses of white, red and other color flowers.
WHERE FOUND: Moist woods, thickets, around homes, schools. Widespread throughout U.S.

BANEBERRY
(DOLL'S EYES)

TOXIC PART: Roots and berries.

SYMPTOMS: Dizziness, increased pulse, vomiting, stomach cramps, circulatory failure, delirium, headache, death.

DESCRIPTION: Grows 1' to 3' high. Toothed leaves 1" to 3" long. Spike-like clusters of small white flowers (May - June).

WHERE FOUND: Woods, moist shaded areas. Not very abundant. Widespread throughout U.S.

CASTOR BEAN (CASTOR OIL PLANT)

TOXIC PART; Entire plant.

SYMPTOMS: Diarrhea, convulsions, excessive thirst, nausea, vomiting, stomach pains, dullness of vision, death in 1 to 12 days.

WARNING: Seeds contain more poison than any other plant part. 1 to 3 seeds fatal to child; 2 to 8 seeds fatal to adults. Seeds may be mistaken for bean-like food.

DESCRIPTION: Shrub-like plant 5' to 12' tall. Star-like leaves. Clusters of inconspicuous flowers (July to September) on plant top.

WHERE FOUND: Common in fields, open woods area, pastures, thickets. Widespread throughout U.S.

CHINABERRY TREE

TOXIC PART: Entire plant.
SYMPTOMS: irregular breathing, paralysis, signs of suffocation, vomiting, nausea, diarrhea.
DESCRIPTION: To 35' tall. Light purple flowers in ball-like masses. Leaves are a natural insecticide.
WHERE FOUND: Thickets, old fields, pastures, etc. Widespread throughout U.S.

DEATH CAMAS

TOXIC PART Entire plant.

SYMPTOMS: Subnormal body temperature, nausea, vomiting, muscular weakness, slow heart beat, diarrhea, stomach pains, death.

DESCRIPTION: Grass-like leaves grow from base. White flowers clustered on top of stem. Green/yellow heart-shaped structure found on petals near flower base.

WARNING: Bulbs of death camas easily confused with wild onions and camas lily bulbs. Remember -- death camas have no onion smell.

WHERE FOUND: Wet or damp meadows, pastures, fields, etc. Dry, rocky slopes. Sunny places. Widespread throughout U.S.

FALSE HELLEBORE

TOXIC PART: Entire plant.
SYMPTOMS: Convulsions, spasms, subnormal body temperature, slow pulse, vomiting, frothing at the mouth, diarrhea, stomach pain, difficult breathing, Fatal if eaten in large quantities.
DESCRIPTION: Grows in dense stands, 5' to 7' high. Heavily veined 8" to 10" pleated leaves. Stout stem. Thick root. Crowded white/greenish star-shaped flowers (June-July) on long branches.
WHERE FOUND: Mountain meadows, wet fields and pastures, swamps. Widespread throughout U.S.

FOXGLOVE

TOXIC PART: Entire plant.

SYMPTOMS: Irregular heartbeat, convulsions, nausea, vomiting, severe headache, tremors, stomach pains, diarrhea, death. Death can occur extremely rapidly.

DESCRIPTION: Tall and erect. Lance-shaped leaves. Stout stem. Showy clusters of thimble-shaped white or colored flowers.

WHERE FOUND: Along roads, logged areas, burned out places. Widespread throughout U.S.

HOLLIES

TOXIC PART: Berries.
SYMPTOMS: Violent vomiting, nausea, stupor, diarrhea.
DESCRIPTION: Evergreen shrub or small tree. Leathery leaves with spine-tipped teeth. Clusters of red or orange berries.
WHERE FOUND: Sandy woods, yards around homes, etc. Widespread throughout U.S.

HORSECHESTNUT

TOXIC PART: Seeds, flowers, leaves, nuts.
SYMPTOMS: Depression, nausea, diarrhea, vomiting, paralysis.
DESCRIPTION: Large tree, palm-shaped leaves. Spikes of showy white flowers (September-October). Thorny husks cover brown nuts.
WHERE FOUND: Often found in yards around homes, parks, etc. Widespread throughout U.S.

IVIES

TOXIC PART: Berries, leaves.
SYMPTOMS: Difficulty breathing, extremely hyper, excitable, profuse sweating, coma.
DESCRIPTION: Hardy creeping plant. Scalloped edged leaves. Violet flowers.
WHERE FOUND: Along sides of roads, lawns. Found climbing on homes, embankments, up trees, etc. Widespread throughout U.S.

JIMSON WEED
(JAMESTOWN WEED, THORNAPPLE)

TOXIC PART: Entire plant.

SYMPTOMS: Hallucination, excessive thirst, dry mouth, vomiting, nausea, delirium, convulsions, coma, death.

DESCRIPTION: Grows 2' to 5'. Purplish branching stems. Trumpet-shaped flowers. Hard, prickly fruit 1" to 2" long. Foul smelling.

WHERE FOUND: Yards, rich farm land, fields, pastures, sides of roads, banks of streams, etc. Widespread throughout U.S.

LANTANA

TOXIC PART: Entire plant.

SYMPTOMS: Collapse of circulatory system. Death.

DESCRIPTION: Shrub-like plant. Strong smelling. Flat-topped colorful flower clusters. Dark blue or black berries.

WHERE FOUND: Along sides of roads, old fields, pastures, etc. Widespread throughout U.S.

LILY-OF-THE-VALLEY

TOXIC PART: Entire plant.
SYMPTOMS: Vomiting, dizziness, stimulation of heart, loss of balance, nausea.
DESCRIPTION: Grows 4' to 8' high. Oval-shaped leaves with smooth edges. Spread by root runners.
WHERE FOUND: Almost everywhere. Widespread throughout U.S.

MONKSHOOD

TOXIC PART: Entire plant.
SYMPTOMS: Convulsions, vomiting, weak pulse, diarrhea, spasms, paralysis of respiratory system, death.
DESCRIPTION: Grows from 2'to 3' high. Buttercup-like leaves. Flowers partly covered by helmet-like growth.
WHERE FOUND: Damp or wet areas. Slopes, woods, thickets, road sides, etc. Widespread throughout U.S.

OLEANDER

TOXIC PART: Entire plant.

SYMPTOMS: Unconsciousness, dizziness, irregular heartbeat, nausea, slowed pulse, vomiting, bloody diarrhea, severe stomach pains, paralysis of lungs, death.

DESCRIPTION: Bush grows to over 18'. Dark green cylindrical leaves. Various flower colors. Brown pod filled with many seeds.

WHERE FOUND: Grown as ornamental plant in yards, playgrounds, parks, etc. Widespread through U.S.

WARNING: Don't cook with oleander wood. Fumes will poison food.

PHYSIC NUT

TOXIC PART: Entire plant.
SYMPTOMS: Stomach cramps, severe diarrhea, convulsions.
DESCRIPTION: Shrub or small tree. Small greenish-yellow flowers. Each apple size fruit has 3 poisonous seeds.
NOTE: Seeds have sweet taste but their oil is violently purgative.
WHERE FOUND: Widespread throughout U.S.

POISON HEMLOCK
(FOOL'S PARSLEY)

TOXIC PART: Entire plant.

SYMPTOMS: Attacks central nervous system. Weakness of muscles, respiratory paralysis, trembling, coldness, coma, death.

DESCRIPTION: Grows to 10'. Smooth, hollow stems with purple spots on lower section. Flat clusters of white flowers (May-August). Turnip-like taproot. Bruised fern-like leaves have obnoxious odor.

WHERE FOUND: Sides of roads, gullies, wet or moist ground, swamps, banks of streams, ditches. Widespread throughout U.S.

WARNING: Don't confuse with Queen Anne's Lace (Wild Carrot). Queen Anne's Lace has carrot smell. Poison Hemlock does not! Queen Anne's Lace has hairy leaves and stems. Poison hemlock does not! A small piece of stem, only 1/2" in diameter, may be fatal.

RHODODENDRON

TOXIC PART: Entire plant.

SYMPTOMS: Blood pressure drops, vomiting, slow pulse, convulsions, watery eyes, runny nose, drooling mouth, paralysis, coma, death.

DESCRIPTION: Small evergreen, 10' to 14' tall. Large, leathery, toothless leaves, rolled under edges. Showy clusters of white or pink spotted flowers (June-July).

WHERE FOUND: Widespread throughout U.S.

ROSARY PEA

TOXIC PART: Seeds.

SYMPTOMS: Circulatory collapse, vomiting, drowsiness, colic, trembling, pulse weak and fast, diarrhea, nausea, cold sweat, stomach pains, weakness, coma, death in 1 to 3 days.

DESCRIPTION: Vine with light purple flowers. Red and black seeds frequently used to make rosaries.

WHERE FOUND: Widespread throughout U.S.

WATER HEMLOCK (SPOTTED COWBANE)

TOXIC PART: Entire plant.

SYMPTOMS: Tremors, violent convulsions, stomach pain, delirium, diarrhea, death.

DESCRIPTION: Grows 2' to 7' high. Stout, bamboo-like hollow stalks. Yellowish oil oozes from roots when cut. Flat spreading clusters of small white flowers.

WHERE FOUND: Ditches, wet or moist ground, swamps, along roadsides. Widespread throughout U.S.

WARNING: One of the most poisonous plants known. Its most poisonous part, the roots, are sometimes mistaken for parsnips.

WISTERIA

TOXIC PART: Seeds.

SYMPTOMS: Stomach pains, diarrhea, nausea, repeated vomiting.

DESCRIPTION: Sturdy vines, smooth bark. Showy clusters of drooping white, lilac or purple flowers. Flat, knobby seed pods.

WHERE FOUND: Wet or damp areas of woods, thickets, pastures, fields. River and stream banks. Widespread throughout U.S.

WARNING: Two seeds are enough to cause serious illness.

YEW

TOXIC PART: Seeds, twigs and leaves.
SYMPTOMS: Difficulty in breathing, trembling, nausea, diarrhea, vomiting, muscular weakness, collapse, coma, death.
DESCRIPTION: Straggly evergreen shrub. Short flat pointed green needles. Twigs smooth. Juicy red berries.
WHERE FOUND: Cool, moist woods. Widespread throughout U.S.
NOTE: Although seeds are poisonous, the red fleshy sweet pulp surrounding the seed is edible in small amounts.

PLANT NOTES

PLANT SKETCH

PLANT NOTES

PLANT SKETCH

PLANT NOTES

PLANT SKETCH

PLANT NOTES

PLANT SKETCH

PLANT NOTES

PLANT SKETCH

PLANT NOTES

PLANT SKETCH

PLANT NOTES

PLANT SKETCH

PLANT NOTES

PLANT SKETCH

THE OFFICIAL
POCKET SURVIVAL
MANUAL
Robert W. Pelton
240
PAGES

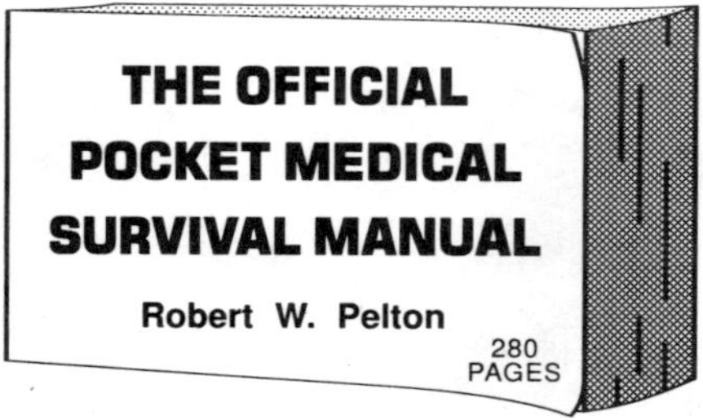
THE OFFICIAL
POCKET MEDICAL
SURVIVAL MANUAL
Robert W. Pelton
280
PAGES

THE OFFICIAL
POCKET MEDICINAL
PLANT SURVIVAL
MANUAL
Robert W. Pelton
280
PAGES

THE OFFICIAL
POCKET EDIBLE
PLANT SURVIVAL
MANUAL
Robert W. Pelton
264
PAGES